NORTH to NUNAVUT

NORTH to NUNAVUT

An Arctic Love Affair

Fred and Joyce Sparling

CHAPEL HILL
PRESS, INC.

FRONT COVER: Loving walrus in Foxe Basin.
BACK COVER: Joyce and Fred in Annie's caribou-skin tent outside Baker Lake.

© 2011 Fred and Joyce Sparling
All rights reserved. No part of this book may be used, reproduced or transmitted in any form or by any means, electronic or mechanical, including photograph, recording, or any information storage or retrieval system, without the express written permission of the author, except where permitted by law.
First Printing

Publisher's Cataloging-In-Publication Data (Prepared by The Donohue Group, Inc.)
Sparling, P. Frederick.
    North to Nunavut : an Arctic love affair / Fred and Joyce Sparling.
    p. : ill., maps ; cm.
    Includes bibliographical references.
    ISBN: 978-1-59715-080-4
    1. Nunavut—Description and travel. 2. Nunavut—Social life and customs. 3. Inuit art. 4. Animals—Nunavut. 5. Animal ecology—Nunavut. 6. Sparling, P. Frederick—Travel—Nunavut. 7. Sparling, Joyce W.—Travel—Nunavut. I. Sparling, Joyce W. II. Title.
F1142 .S62 2011
917.195   2011929914

*This book was written for our four children,
their spouses, and our ten grandchildren.*

*We thank you for your love and support of our
passions, and know that you will continue to
respect the natural world and all its people.*

CONTENTS

PROLOGUE		ix
CHAPTER 1	A New Passion: Inuit Art	1
CHAPTER 2	The Way North: Greenland to Nunavut	11
CHAPTER 3	The High Arctic: Resolute to Greenland	27
CHAPTER 4	Where the Bull Caribou Gather: Pangnirtung	41
CHAPTER 5	An Unusual Fortieth: Arviat and Baker Lake	55
CHAPTER 6	A Secret River: To the Thelon River	81
CHAPTER 7	Golden Caribou: Baker Lake and the Kazan River	103
CHAPTER 8	Wild River: To the Back River	135
CHAPTER 9	Forty Below: Igloolik	149
CHAPTER 10	Tundra Camp: Baker Lake	165
CHAPTER 11	August Ice: Igloolik	181
CHAPTER 12	Friends of the Family: A Return to Igloolik	203
CHAPTER 13	Reflections	213

ACKNOWLEDGMENTS 217

NOTES 219

BIBLIOGRAPHY 229

GLOSSARY 235

MAPS

MAP 1 Arctic Canada and Greenland 2
MAP 2 Thelon River 82
MAP 3 Kazan River 104
MAP 4 Back River 136
MAP 5 Igloolik—Foxe Basin 150

PROLOGUE

Our story is a memoir of ten years of traveling in the Canadian Arctic in our late sixties and early seventies, and a reflection on what we learned in the process. Our journeys were at an age when such adventures seemed unlikely, emphasizing possibilities rather than limitations.

The odyssey started with our rediscovery of Inuit art. The art fired our imaginations and raised our curiosity about the people who created it. Touring on ships through the North provided an introduction to the people of the eastern Canadian Arctic and an overview of their beautiful land, Nunavut. Soon a view from afar was not enough, and we embarked on a quest to get onto the land and learn about the people. Canoe trips through the Barren Grounds, across vast lakes, and down turbulent rivers took us to places vacated only recently by the Inuit, and still crucial to their sense of self. Multiple visits with Inuit in several hamlets enabled connections to the people. We became friends with Inuit families, moving from superficial to deeper understanding, from acquaintance to personal relationships. Correspondence by email, only possible in recent years through the miracles of satellites, computers, and generator-derived electric power, added depth and nuance to these friendships. The continuous thread throughout these travels is our ever-growing appreciation of the Inuit. They have lived in the Arctic for a long time and have been shaped by the cold, the beauty, and the difficulties of living in extreme conditions. The Inuit have much to teach.

In sharing our experiences, we hope others, especially our children and grandchildren, will come to know and understand the people of the far North, and to appreciate the importance and magnificence of their wild lands. Already our Inuit friend Jeena and our daughter Betsey have conversed via email, and our grandchildren have received positive comments from Jeena, who lives near Santa, about receiving Christmas gifts.

The animals of the Arctic are under threat from a multitude of forces not under the control of the Inuit, including global warming and habitat destruction. We saw caribou, white wolves, musk-oxen, seals, walrus, and polar bears up very close, and we were changed by the experience. This book is a plea to help save this great wilderness before it is too late.

Throughout our travels, we each recorded in written notes our conversations on the day on which they occurred. Tape recorders were not used. With two reasonably coherent recorders, the quotes are as close as possible to the exact words spoken. Our Inuit friends have agreed that we have not erred too egregiously in retelling their stories, and they have given permission for sharing their anecdotes.

1

A NEW PASSION: INUIT ART

What I love and collect is that which evokes responses in me.
GEORGE SWINTON[1]

SEPTEMBER 2000 It all started with a bear. Not any bear, just that one bear. If it hadn't been for the bear encounter, it never would have happened. Somehow, that lustrous, dark green, black-speckled, long-necked bear with a pendulous belly changed and energized us, accessing veins that ran deep and long. Seeds of this new passion had been planted many years before, but they were dormant. A catalyst was needed to activate them. The bear lit a fire and opened our eyes to new opportunities, new vistas.

It was September, a warm day in Toronto, and I (Fred) was scheduled to deliver a scientific talk in the early afternoon. The meeting seemed tiresome, but it was undoubtedly I who was tired after a long career as a physician, scientist, and administrator. A leisurely three-mile walk to the convention center brought me by an Inuit art gallery. Inside were hundreds of small to medium-sized carvings of animals and birds, drum dancers, hunters, women carrying children, and more, cluttered on tabletops and shelves and in window displays. All were made by the eastern Canadian Inuit, from the

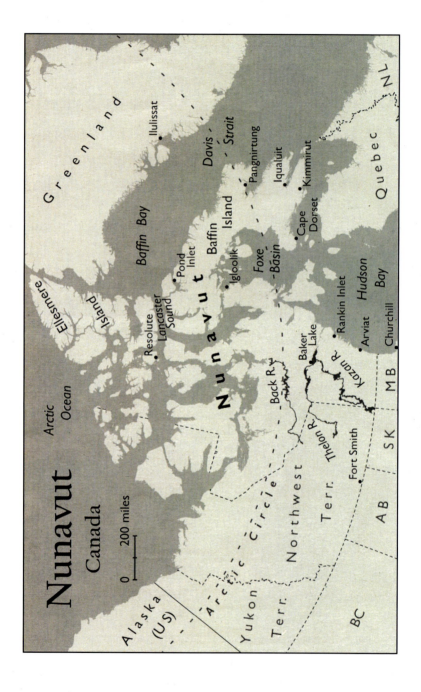

land recently designated Nunavut. Many of these works were repetitious variations on a few themes, but some had real individuality and character. A quality pyramid was readily evident.

One bear seemed weird, and it held my attention for some reason. Resisting temptation, I ambled down to the convention center and gave my talk. Later, back in the hotel, it was not science that lit my thoughts, but a two-foot-long, eighty-pound stone bear. I called home and told Joyce I had fallen in love with a bear. She delights in retelling the story, saying, "Fred called home and said he had fallen in love," omitting any mention of the final phrase "with a bear." Despite her playful twisting of this discussion, Joyce told me to follow my heart. Relieved of inhibitions, I took in another gallery and quickly developed a fatal attraction for another large stone beast, this time a figure of a musk-ox in light grayish-green stone with carved caribou antler horns. These pieces were emblematic of a culture that seemed exotic, the products of unknown ambassadors from the North. Fortunately, Joyce already knew and cared about Inuit art. Curiously, we observed few other couples for whom a similar passion was shared equally. So started an adventure that grew to be much larger than collecting stone carvings.

We launched an extensive search of the literature in an effort to understand the people and the art. Swinton's book, *Sculpture of the Inuit*, was for us and most collectors a sort of unofficial Bible of Inuit art. Visits to galleries in the United States and Canada, and Canadian museums, helped us to learn what is good and what is merely pretty and glossy.[2] We came to understand what one collector friend told us—"This is blood sport"—when we were forced to drop out from the bidding at the annual November Waddington's Inuit art auction in Toronto. Other collectors will readily understand this learning curve, and the pleasures, frustrations, joys, and disappointments of collecting. Our appreciation grew for the abstract work of the greatest artists from the western shores of Hudson Bay in the tiny hamlets of Arviat[3] and Baker Lake and Rankin,[4] and for the fabulous embroidered works on cloth made by the women of Baker Lake.[5] The collection became a living, changing thing.

FIGURE 1 John Tiktak, *Head*, c. 1970, Rankin Inlet. Simplicity, essence.

FIGURE 2 John Kavik, Old Man, c. 1975, Rankin Inlet.

James Houston was an important figure in the story of Inuit art. In the late 1940s he traveled to and then chose to live for many years with his family in the eastern Canadian Arctic, helping to stimulate and support the creation of marketable carvings and print art among the peoples of the eastern shore of Hudson Bay and Cape Dorset.[6] His influence extended to all of the Inuit communities of eastern and central Canada; absence of similar art among the Inuit of Alaska and Greenland testifies to the power of his influence. His role was crucial in the 1950s and 1960s when Inuit were being moved from hunting camps on the land to money economies in government-organized hamlets. The art grew out of their tradition of carving small amulets and sewing animal-skin clothing. Carvings and prints brought in much-needed money, but they also were expressions of their angst about the increasingly rapid loss of their ancient hunter-gatherer culture.

Many Inuit artists understandably were motivated by commercialism, forcing the repetition of figures and forms attractive to southern consumers. At the bottom end, the work was of souvenir quality. Personal pride and individuality, however, drove the best carvers to create work that was original and could be designated as art. The quality of the best pieces was astounding, considering the lack of tools other than axes, files, and sandpaper, and the harsh conditions under which the artists worked. At the top end of the genre, we could identify individual artists whose work was moving and expressive, possessed of interior fire and feeling. Some of these works were raw and stark, others smooth and flowing, often laced with humor and self-deprecation—sometimes in rough gray stone, other times in lustrous stone that polished to a gloss, with deep and varied hues of green, blue, brown, and many other colors. Carved caribou pieces carried the memory of the tundra with them. Old ivory pieces were imbued with a depth of color, and whalebone pieces with unusual form and texture. The prints, drawings, and wall hangings were full of color and fabulous imagery. Some of these graphically showed the old ways of life or old oral traditions or taboos, whereas others were abstractions of the tundra or of the human condition.

FIGURE 3 Jesse Oonark, *Untitled*, 1978, Baker Lake. One side of double-sided wall hanging. May depict her varied spiritual beliefs of shamanism and Christianity: Christian figures (clerics?) in black and red in center; and (twelve apostles?) in blue and pink on outside margin; central caribou-shaman figures in blue.

FIGURE 4 Jesse Oonark, *Untitled*, 1978, Baker Lake. Reverse side of above wall hanging. An Inuk woman with ulus (woman's knife) radiating from a central face; a self portrait?

But who were the people who did this marvelous work? Our knowledge of the Inuit was very small. Growing up we knew them as Eskimos, allegedly an old name for people who eat raw flesh. Eskimoan peoples are known by many names from Siberia to Greenland; in Alaska they are either Yupit or Inupiat, but in central and eastern Canada they are Inuit, "the people." For over two centuries they had interacted with whalers along the coastlines of Hudson Bay and Baffin Island, and intermittently for at least a century or two longer with Hudson Bay traders.[7] Missionaries had been in or near most communities for a century or more. Almost all Inuit had been contacted by European peoples in some way by the middle of the nineteenth century. Nevertheless, Inuit were not well-known to most Americans. The inland Inuit of the central Canadian Barren Grounds were discovered by most people after Farley Mowat published his widely read book in 1952.[8] People were shocked by tales of the starvations of the most southerly Inuit group, the Ahiarmiut, and the management of their communities by well-intentioned but ill-informed white government bureaucrats. In 1952 Richard Harrington published descriptions of the terrible starvations among the Paallirmiut (Padlermiut), who lived to the east of the Ahiarmiut.[9] His photos became icons—beautiful in their stark portrayals of human dignity in the face of tragedy. Geert van den Steenhoven described in a prospective report from 1955 (published later in 1968) the life of these inland nomadic people[10] just prior to their displacement to another site: a catastrophic move to a place with poor hunting, where suffering increased. To our surprise, these aboriginal hunter-gatherers were living just to our north. Our notion of them was romantic, and not entirely correct. The Inuit had been living in the North for millennia, and some of the Inuit were more Westernized than southerners' idealized images portrayed.

Despite our ignorance of them, Eskimoan people arrived in the North five thousand years ago, according to the archaeologist and scholar Robert McGhee.[11] Indians came several thousand or more years earlier, and remained at the tree line or below. Eskimoans stayed within the Arctic, migrating from Asia in several distinct waves across the top of North America from west to east, from Alaska to Greenland, following animals and searching for sources of iron. The first to come were people of the Arctic Small Tool tradition (ASTt), whose small camps were scattered across the high Arctic. They evidently perished. In McGhee's formulation, the subsequent Eskimoan immigrants can be divided into essentially two major groups, not closely related to each other: the Dorset and the Thule. The Dorset period of dominance lasted from about thirty-five hundred to four thousand years ago to about one thousand years ago. The early Dorset, also called Paleoeskimos, spread throughout the Arctic and created abundant small yet lovely ivory carvings. They built snow houses as well as partly subterranean homes, and hunted seals, but not whales.

For unknown reasons, the Thule culture replaced the Dorset, starting about one thousand years ago at a time of relative climate warming and probable changes in the abundance of their preferred food sources. The Thule were a physically smaller but more aggressive, less artistic people, with a well-developed whale-hunting culture that included sinew-strengthened recurved bows that were more powerful than the simpler bows of their predecessors. They constructed whaling boats (*umiaq*) and improved kayaks (*qajaq*). Where bowhead whales were abundant, they built substantial communities, utilizing whalebones as part of their architecture. Their migrations may have been influenced by knowledge of European colonizers and migrant fishermen on the eastern coasts of North America and Greenland, who provided a source of iron and other trading goods.[11] From the Dorset, the Thule may have learned certain cultural skills, including making snow houses. Some Dorset apparently survived on Southampton Island at the north end of Hudson Bay until 1902, when they were wiped out by an epidemic. Modern Inuit referred to earlier people as Tuniit, whom they characterized as giants; scholars now believe that the Tuniit were the Dorset.[12] Inuit art often shows images in which a giant is slain, possibly reflecting ancient oral histories.

The Thule probably evolved gradually into modern Inuit, coincident with the onset of much colder weather again about five hundred years ago (the

Little Ice Age). The Little Ice Age lasted until the mid-nineteenth century. The increased cold forced the Thule to abandon their permanent whaling camps. They became more migratory, living in igloos and skin tents and following fish and caribou. This latter culture is what we identify as Inuit. Many Inuit migrated south, including those from the Coronation Gulf area in central northern Canada who moved southeast, establishing inland Inuit communities.[12] All of the Inuit communities were small, organized in widely dispersed hunting camps. Except for the inland communities of the Caribou Inuit, all had a base on the sea, where seals and walrus were plentiful. All Inuit followed migratory caribou herds, and hunted abundant fish, birds and bird eggs, and occasionally bear or musk-oxen. A complex set of taboos and spiritual guidance by shamans ruled much of their lives. Periodic starvations plagued them, but infectious diseases were unknown before the arrival of white men.

We were oblivious to all of this as we finished high school and entered university. At the time of the great civil rights changes in the United States in the 1960s, while we were finishing our professional training, marrying, and starting a family, the Inuit of the Northwestern Territories were being relocated into government-subsidized and -administered communities where starvation could be prevented, medical care provided, and schooling and government aid administered. Experiments were undertaken, relocating Inuit into the high Arctic in places where Inuit had never lived in modern times.[13] Moves off the land to organized communities brought radical change to the structure of their society. Dog teams were replaced by snowmobiles; igloos and caribou-skin tents were replaced by prefabricated heated houses; skin clothing was replaced by store-bought clothing. Because white people (*Qallunaat*)[14] had trouble keeping track of Inuit, each person was assigned an E (for east) or W (for west) number, which they were to wear around their necks. Later this practice was replaced by "Operation Surname," in which Inuit were required to choose a last name. Brothers often chose different last names, resulting in some confusion about relationships. For collectors of Inuit art, the E numbers did have the virtue of clarifying the identities of individual artists.

At the time of the relocations there may have been only ten thousand to twelve thousand Inuit. Their numbers are increasing rapidly due to better health, and roughly thirty-five thousand modern Inuit now live in eastern and central Canada. They are connected to the outside world by generator-derived electric power and satellite dishes aimed low on the horizon, bringing the miracles of television and computers. Oral histories and telling of old legends have been substantially supplanted by these new technologies. Knowledge of old survival skills are said to be fading. Parental control and influence have been replaced in many ways by the authorities of white schoolteachers, the Royal Canadian Mounted Police (RCMP), and priests and pastors. Depression, suicide, drug abuse, and domestic violence have replaced starvation as problems in the communities. Yet as we have come to know, the people remain generous and friendly, despite intrusions from the culture of the South.

In collecting Inuit carvings, prints, drawings, and wall hangings, we were communicating with their culture. We became frustrated by our lack of understanding of the art within the Inuit context. How could we travel to their isolated home? There are no roads to the Inuit lands of Nunavut and Nunavik. Occupying the eastern half of the old Northwest Territories, Nunavut became a new Canadian territory on April 1, 1999.[15] Nunavik, previously known as Arctic Quebec, lies south of Nunavut across Hudson Strait, and is part of the province of Quebec. We wanted to experience these lands, to see firsthand where the art was made, and to communicate with these enduring and talented people. It was getting late for us, and perhaps for them, too, with the continuing onslaught of Western television and material goods, as well as global warming threatening their entire culture and ecosystem.[16]

2

THE WAY NORTH: GREENLAND TO NUNAVUT

*All the words in the world could never have expressed
for me the secret of the charm of that land.*

WINIFRED PETCHEY MARSH[1]

JULY 2001 The advice and generous friendship of H. G. Jones, a local collector of Inuit art and long-experienced traveler to the Inuit communities, showed us the way north. In late 2000 he told us there might be a slot on a July 2001 Adventure Canada cruise that was scheduled to travel from Greenland to Hudson Bay, with stops in important Inuit art communities including Cape Dorset on southern Baffin Island. We called, found there was room, and committed to go.

The literature of exploration of the North is crammed with stories of intense cold, scurvy, starvation, being cast adrift on ice flows, and extraordinary feats of endurance.[2] The wisest of the explorers lived with the Inuit and learned to adapt to their ways.[3] Those who insisted on doing things as they had always been done often paid for their arrogance and stubbornness with their or their men's lives. We were going in an entirely different era, when tourists such as we could see the same sights as the early explorers, but in comfort and safety. Nevertheless, we were full of anticipatory awe.

Discovered by the Western world in the late tenth century by the famous Viking, Erik the Red, Greenland is a massive island covered up to two miles deep over most of its surface by a huge ice cap. The Norse colony on the southwest coast of Greenland persisted and thrived for several hundred years, but mysteriously disappeared in the early fifteenth century, perhaps because of the onset of colder weather in the Little Ice Age.[4] The first European to visit again was John Davis in the late sixteenth century; the straits between Greenland and Canada were named Davis Strait in recognition of his pioneering exploration. There was a lot of history in the waters between Greenland and Canada. Armed with cameras on a strong and stable ship equipped with electronic navigation devices that Davis could not possibly have imagined, we cruised comfortably on some of the same waters.

A chartered jet carried about one hundred of us from Ottawa to an old U.S. Cold War air force base at Kangerlussuaq (formerly Sondre Stromfjord), below the Arctic Circle on the west coast of Greenland. As we passed the Canadian coast, the seas below were heavily laden with massive chunks of flat pan ice, far into Davis Strait. In a seeming blink, we arrived at a land that was barely green. Kangerlussuaq has nothing special to recommend it other than its proximity to the fjord, and its handling of large jets. We and our gear were taken to Zodiacs (small rubber boats) at the water's edge, and motored across milky, greenish-white, glacial, silt-laden waters to the *Akademik Ioffe*, our home for eleven days. This ice-strengthened Russian ship was built to perform acoustic research, but after the end of the Cold War was no longer needed to monitor U.S. submarines amid the polar ice.

Rockwell Kent wrote of his time in Greenland in the 1920s: "A whole new world of land and sea rises to meet me...far snowy peaks and dazzling glimpses of the inland ice, mountains and headlands, islands, bays and inlets; and the ocean—blue and calm. Greenland! Oh God, how beautiful the world can be!"[5] As we cruised down the very long Kangerlussuaq Fjord toward Davis Strait, the light at midnight was soft yet somehow intense, and we bundled up to fight off the cold winds on the upper decks.

Our companions on the staff included two Inuit, Aaju Peter and Pakak Innukshuk. Aaju was Greenlandic Inuit but now lived in Iqaluit on Baffin Island; she was about to enter the first law school class in Iqaluit. Pakak was from Igloolik, an island at the top end of Foxe Basin, near the entry to Fury and Hecla Strait. Both of these were places we would come to know. Pakak was famous for his role in the Cannes Film Festival–winning film *Fast Runner*, made by an all-Inuit company in Igloolik. The staff also included John Houston, son of James Houston of Inuit art fame. John grew up in Cape Dorset, and he knew well many of the great artists in Cape Dorset. Trip leader Laurie Dexter had been an Anglican minister in Pond Inlet at the north end of Baffin Island for a decade. Both Laurie and John were fluent in Inuktitut. Laurie had skied from Russia to Canada across the North Pole with a small band of Canadians and Russians, an amazing feat. The Adventure Canada trip was tame for someone with his credentials, but we were glad to have him out front when decisions were made.

Before long, we crossed the Arctic Circle, on our way to Disko Bay and Ilulissat, formerly called Jacobshavn. There was magic in the mere crossing of the Arctic Circle, as though we had entered a new realm of some sort, a land of mystery and awe. We have never felt the same about the Antarctic, even though we knew that fearsome place was colder and even more inhospitable. Why was that? Perhaps it was because the Arctic is in fact more accessible and livable at its margins. Indeed a great measure of the attraction of the Arctic is that it has been inhabited by hardy and ingenious aboriginal peoples for at least four millennia. Explorers have come here for centuries, much longer than to the Antarctic, and their stories fascinate us. Explorers from Europe may actually have ventured up here since the time of the ancient Greeks, if we are to believe the stories attributed to Pytheas, a fourth-century BC Greek navigator and northern adventurer.[6] Tales of unicorns, mythological horselike creatures, might have been inspired by observations of narwhals with their long single spiral horn. Somehow, images of the North have entered our collective consciousness. The lure of the North runs deep. We were merely latecomers to an old story.

Icebergs were small and white on the horizon as we approached Disko Bay. Soon they looked a lot bigger; there were many more of them, in all sorts of fantastic shapes—some huge with a jagged dirty gray surface, others softly contoured and smooth, in various shades of white and greenish blue, or a light gray-blue. Depending on the light, they sometimes were deep blue, and even golden. The icebergs seemed to generate their own wisps of fog. The smoothed and shaped bergs had been at sea longer and had melted and rolled over many times, creating a ridged and sculptured look. In some there were graceful open arches, in others vertical walls where a large chunk had calved. Under an overcast sky, with beams of bright light peering through, they were beautiful. Some of them—indeed, many of them—were huge, about two hundred feet above water, meaning their total vertical dimension was at least fifteen hundred feet. These were real titans. A large coastal ferry passed very close by one, which enabled us to gauge its size; the massive, flat-topped, broad iceberg was at least twice the height of the ferry. The bergs moved very slowly, pulled along by deep currents heading north up the coast, paradoxically sailing into the wind at times, before sweeping west and then down the east coast of Canada past Baffin Island, Labrador, and Newfoundland, until the bergs eventually melted away in the North Atlantic.

Whalers in the great days of sailing ships had to contend with these monster bergs, a task difficult because of the vagaries of winds and currents. Even we in our much more maneuverable and strong ship had to be aware of submerged, smaller, yet still dangerous "growlers" and pieces of old bergs. A ship similar to ours, the famous little red ship *Explorer*, was sunk in 2005 by an encounter with a submerged iceberg in the Antarctic. While we were contemplating such concerns, a bright red fishing trawler, not very large, emerged from the fog, passing quickly through the ice pans.

FIGURE 5 Icebergs encountered off Disko Bay, Greenland.

Seeing icebergs from the upper decks of a ship is one thing, and passing alongside in their shadow at water's edge is quite another experience. We debarked down a narrow gangplank to the Zodiacs and cruised slowly among the bergs, in absolutely calm water. A few gulls flashed by, but otherwise the waters seemed devoid of life, empty except for the overwhelming presence of the bergs. We passed very close to a huge, light-gray-and-blue behemoth, the highest of the group with a sheer cliff face towering above us, which alarmed Aaju greatly. "We should not go so close. If it calves we will be swamped and killed. Get away." Nothing happened, but she was right. Icebergs are dangerous, not only because falling pieces or a sudden rollover can create huge waves, but also because large pieces may break loose underwater, bursting through the surface like a missile with no warning and disastrous consequences for those in the neighborhood. It was, however, a photographer's heaven. The sea was various shades of gray, with white and azure icebergs casting perfect dark reflections onto the water. There were narrow passes between some of them, leading to vistas seen only dimly. Some had caves carved by the erosion of the water. Floating quietly without the noise of the outboard motor, the experience was sensual, a beautiful, peaceful dream—a magical place where fantasy and reality merged.

Dozens of red or yellow or blue small wooden or fiberglass skiffs dotted the harbor in Ilulissat. These boats appeared to be a modern cross between a kayak and an *umiaq*, a large skin-covered boat generally paddled by women. Racks of modern kayaks were stacked by the edge of the bay, brightly colored sleek, all plastic craft. The old skin kayaks and umiaq had no place in the modern world. The Danish influence was evident; houses of bright yellow, blue, and green dotted the hillside. Stores contained ivory totems (*tupilaks*), sealskin, and rabbit clothing; there was at least one coffee shop and a little restaurant, as well as fish processing plants and the necessities of a modern small town.

The home of the famous explorer Knud Rasmussen was now a small museum. His little red home with steep roof sat next to a charming red church; these buildings were set amid green grass and fields of white Arctic cotton, and against a background of icebergs on a foggy sea. Knud was one-eighth Inuit, and he was raised both as a Dane and an Inuk. Determined to learn of his Inuit roots, he spent many years among the Inuit of northwest Greenland, where he learned to speak fluent Inuktitut. His explorations included the far northern wastelands of Greenland, where he experienced starvations and death of some of his companions, but his greatest impact came from his epic three-year dogsled journey across Arctic Canada from 1921 to 1924 (the Fifth Thule Expedition). Rasmussen traveled all the way to Alaska by dog team with an Inuit woman, who not only sewed his clothes but bore him a child. A shaman predicted he would travel safely: they did survive without major mishap on a journey fully the equal of the Lewis and Clark expeditions in the western United States in the early nineteenth century. His aim was neither to map nor to conquer, but rather to record the customs and beliefs of the people before they were lost by the incursions of southern society. Rasmussen observed[7] ancestors of the people we were to meet later in Igloolik and in the

Barren Grounds of Canada. His Greenlandic Inuktitut was understood all the way to Alaska, and many of the stories and beliefs he learned in northern Greenland were shared as well, testimony to the common origins of the small bands of people now so widely scattered.

Outside of town, we walked across barren rocky fields that led to the Ilulissat icefjord, which runs twenty-five miles from the edge of the great Jacobshavn Isbrae glacier, down past Ilulissat, emptying into Disko Bay. Joyce stopped for a moment, standing alone, taking in our first sighting of massive bergs lined up in the frozen fjord below us, their white peaks rising like small mountains into a low ceiling of fog. All was white, and silent. "I have come to heaven," she said. Her reverie was interrupted twice when we heard a thunderous sound from up the fjord, as new bergs separated from the glacier. We walked on down, passing some very old Dorset culture homes and graves, perhaps two thousand years old. Several human skulls and other bones were well preserved in a small but deep hole in a cluster of rough rocks. Once there had been a community here. Now it was silent witness to one of the greatest birthplaces of icebergs on earth, each torn from the face of the fast-moving glacier. The great bergs often are trapped as they move majestically down the ice river to the sea, waiting for the right tide and wind to get over a barrier lip before starting their long journey north in the Greenland current. We could spend a month here among the icebergs, catching them in various attitudes and light and color, and never be bored.

Back on ship, we departed amid the ice, past dark cliffs under low-lying clouds and fog, with a thin line of light on the horizon. It was a scene we had seen in old paintings, prints, and drawings from the early days of exploration in the Arctic. Baffin Island and our destination of Cumberland Sound lay four hundred miles west, across Davis Strait. There were a few pilot whales, some icebergs with deep blue-green caves carved out by the erosion of the water, and an occasional seal sunning on bits of pan ice. This was a time for contemplation in the cool air, and for reading to the accompaniment of the steady thrumming of the engines and the gentle rolling of the ship.

Unfortunately, pack ice was massed by wind and tide outside Cumberland Sound to a distance of sixty miles, in layers up to sixteen feet high in places. We were awed by the thoughts of the endurance and strength required by the Inuit and the explorers to get through such tortured pressure ridges hauling heavy sledges, even with modern snowmobiles. The ice forced us to abandon plans to visit the picturesque hamlet of Pangnirtung, deep in Cumberland Sound. We headed directly south for Hudson Strait and the Inuit communities of Kimmirut and Cape Dorset.

With the released time we made an unscheduled stop at the Savage Islands, just at the entrance to Hudson Strait. Henry Hudson and his men sailed through this turbulent strait in 1610–1611 and discovered a great bay, but were beset by ice and starvation. Both the strait and the bay were named for him even though his men famously mutinied and cast him adrift to die. Many others subsequently sailed into Hudson Strait and across Hudson Bay in a fruitless search for a Northwest Passage.[8,9]

The Zodiacs were dropped well outside the rip between the islands, which were partially obscured in fog. We passed between bare rock islands with snow lying in crevices, cold and bleak in every way, seeking the polar bears that were often here at this time of year. A very large bear prowled the rough gray rocks close to the shore. We approached cautiously, realizing how small and vulnerable we were in our relatively tiny, soft rubber Zodiac. The bear ambled over toward us, sniffing the air, angling his long neck up to catch our scent, obviously nervous, but acting the king that he was, and reminding us of our first sculpture resting safely at home. There was nowhere for this bear to hide, but certainly he was not going to be moved from his spot. He was almost pure white, with blue-black nose and black eyes, and powerful shoulders and rump, dignified and yet scary. Deeper into the pass, with the big male safely distant, we clambered ashore, looking at the beautiful small alpine flowers at our feet. Pakak and John Houston carried rifles, acting as guards, and too soon we heard commands to get back aboard the Zodiacs immediately because a mother bear (*atirtalik*) and two cubs (*atirtaq*) were approaching over a ridge. Very dense fog suddenly swept down on us as we returned to the ship. We spied the mother and her two nearly full-grown cubs in the water, swimming to get away from us. This was bear heaven! In summer when the ice is largely gone, these rocky islands offer a superb refuge and seal hunting for the bears.

On the south coast of Baffin Island, we sailed into Kimmirut (formerly Lake Harbour), an attractive little hamlet of several hundred people set at the back of a small fjord. The hills above town were a soft golden brown and utterly barren, except for a white Hudson's Bay Company (HBC) sign made of white stone embedded in the turf. The sign was a reminder of the early history of the Canadian North, indeed of all Canada, because the HBC had opened up the country literally in its quest for riches from the fur trade. Small prefabricated houses in dull grays and whites were crowded together close to shore. Again the Zodiacs delivered us, and we clambered onto the gravel and sand to be greeted by Inuit hosts or hostesses, patiently waiting for us. We joined a group led by an attractive young woman, Annie Qimirpik, accompanied by her four-year-old daughter. Each was in a cloth *amautik* (woman's parka with hood); a doll peered from the hood of the appealing daughter's amautik. Annie spoke softly, and moved slowly; she was graceful in her speech and manners, making us feel a bit clumsy somehow. We made our way slowly to the town, although we were anxious to see the available carvings before others scooped up the best. We tried to hold back lest we betray our competitive, crude, materialistic southern manners.

Yes, Annie knew of Nujaliaq Qimirpik, who carved the beautiful pale green and yellow-brown striped musk-ox, our second piece and friend of the bear that initiated our collection. He was her father. No, he was not in town, but she was a carver also; she liked to carve mother-child figures. Annie walked us to her father's home on a lovely site overlooking the harbor, where her brother was working on a stone bear next to his carving shed. He smiled broadly, showing a totally toothless mouth, but a very happy and open face. Listening to them speak together, we were fascinated by the inflections and rhythms of Inuktitut that flavored and softened their language, adding a musical quality, sounds we loved the first time we heard them. The polysyllabic language is full of very long compound words, adding shades of meaning, but confounding the would-be student of the language.

We asked about Elijah Michael, who carved a large piece in our collection, depicting a dancing shaman transforming into animals (owl, caribou, and walrus). Traditional Inuit belief held that we are one with the animals on which we depend, and that animals give themselves to us so that we can survive. Shamans had the power to transform into animals, as well as to undertake spirit journeys to the moon or deep into the sea. Annie guided us to a small shed where Elijah was showing his work. A gray, unpolished stone bear was lying on the ground outside, surrounded by files and a hatchet. Inside we found Elijah with another well-known carver, Iola Ikkidluak. We tried to talk to Elijah, but he spoke no English, and Annie had disappeared. He watched patiently as Fred tried to pantomime our carving of a shaman transforming into animals, but Elijah only laughed, no doubt at Fred for his silliness. On the table in front of Iola, a strong bearlike man, was a green stone carving of a shaman turning into a bear. His charming and attractive teenaged daughter told us this depicted the story of a man who became so angry that he turned into a bear, killing his wife. She also told us he would rather carve figures other than bears, but he is forced to carve what people will buy. Years ago he carved some highly imaginative pieces, now included in museum collections.

Not many ships came to Kimmirut, so the whole town was out to greet us. Smiling kids and adults were everywhere, welcoming and friendly. The atmosphere was relaxed, giving the hamlet and its people an aura of purity and innocence. It would be good to return; there was a little hotel and an inn where we could stay. A plane made scheduled flights from Iqaluit, presumably landing in the harbor in summer or on the ice in winter. Perhaps we could canoe from the highlands down the river that came right into the hamlet. Guided canoe trips could be arranged. Sometimes dreams happen.

Rose-colored skies reflected on the dark waters as we departed Kimmirut. Avoiding shipboard parties, we were more than content to be in the Arctic and meet the Inuit. Maybe the problem was in us, since Knud Rasmussen and his companion Peter Freuchen were famous party givers always on the lookout for a celebration.[10] After crossing to northern Quebec, we headed back north across Hudson Strait to Baffin Island and Cape Dorset (Kinngait), the most famous Inuit art community.

To some collectors of Inuit art, Cape Dorset is a holy land. It was here almost sixty years ago that James Houston encouraged and stimulated carving in stone, and taught the Inuit to make prints on paper. Lovely soft green and marbled serpentine and beautiful white or pink marble were found in three locations[11] many miles east along the southern shores of Baffin Island, halfway to Kimmirut. The serpentine was just the right composition to allow detailed and intricate carvings to be made, with open ("negative") space and extended wings or extremities. The Inuit mined the stone by hand and transported it back to Cape Dorset in summer either by small boats or canoes, or in the winter by qamutiik over the ice. In the early days the carving was done by simple hand tools, but later with power tools. To avoid inhalation of stone dust, the carvers often worked outside in the cold. The stone also was used for carving print stones. Prints made from these stones were influenced heavily by Japanese traditions and techniques, learned by Houston during a sort of sabbatical leave.

The community of Cape Dorset was under low clouds on a cool and gray day. Several famous artists from Cape Dorset, including Kenojuak Ashevak, joined us for lunch. There was little chance to talk, although it was moving to meet such an iconic figure. There was no Annie Qimirpik to guide us, no personal touch. We realized it would take a longer visit to have meaningful conversations. Many white visitors have stopped here over the years, and visiting Qallunaat are not a novelty. There were rumors of a recent physical assault in the home of one of the prominent artists, and that may have dampened the mood. Perhaps we expected too much. We anticipated learning of

Cape Dorset artists' development and uniqueness, but one can't see creativity by walking the dirt streets and visiting the schools and community centers. Aaju had the right idea. She went about town on her own, going into homes of people whom she did not know, seeking—and finding—country food (caribou, char, seal, whale skin).

A memorial to Peter Pitseolak above town offered a view of the community framed by low mountains. Pitseolak is celebrated primarily for his leadership in the community, and for his photographs that documented the old ways of life that were rapidly dissipating under the onslaught of southern culture,[12] but he was a talented artist as well. We have in our collection a large piece in old apple-green serpentine he carved over forty years ago. His Sedna on narwhal carries a description in syllabic Inuktitut of the Sedna myth on the base of the mermaid Sedna figure and the narwhal on which she stands. Sedna was an often angry goddess figure whose moods controlled the abundance of the animals on which the hunters depended. When she was angry, a shaman (*angakkuq*) might make a journey to the bottom of the sea to placate her, entering her lair and combing her hair. Then the animals would return. One expert declared, "This is Sedna carved by someone who believed in her." Once she ruled absolutely among the Inuit; now she rules our living room, her enigmatic Mona Lisa–like face gazing silently at us as we go about our lives.

Across from Cape Dorset lies Mallikjuaq Island, an uninhabited old Thule culture site. Perhaps a dozen sunken, stone-lined, tiny homes were clustered together. They once were covered with whalebone roofs; scattered whalebones were still in evidence. Entrance tunnels remained intact in some. Each home had a cooking hearth and a stone sleeping platform, and we sat in one of them talking quietly, thinking about life here five hundred or more years ago. Pakak picked up his caribou skin drum and danced amid the ruins, bobbing slowly, chanting and calling out his song, framed by dwarf Arctic fireweed at his feet and a cold blue sea in the background. A ptarmigan was beautifully camouflaged among the stones and gravel, invisible in full sight. One stone inuksuk, a stone statue mimicking a man, stood silently. Cape

Dorset could be seen across the bay, partially hidden in persistent fog and gray mist. A beautiful natural rock garden could not be entirely avoided as we walked down an inclined beach, marked by a discrete series of flat terraces. The land was still rising after the melting ten thousand years ago of the great ice sheets that crushed the land for so long.

A nocturnal cruise brought us to a small, mist-shrouded, bare rock island far from land out in northern Hudson Bay. Walrus were gathered on one pole of the island, looking cold and gray in the water but a warm tan on the sunny rocks. Bulky bodies were heaped on each other, many with remarkably long tusks. A quick circumnavigation in Zodiacs revealed on the back side of the island a large polar bear covered in blood, glowering over a walrus that he killed and was in the midst of devouring. This still is a wild place.

In Coral Harbor, on Southampton Island, at the top end of Hudson Bay, a memorable fashion show of children in caribou-skin clothing was held for us in the school gymnasium. The children's often rotund mothers beamed with happiness at their children and at their own expert sewing of these outfits, perhaps sewn by the mothers in a burst of community pride that accompanied the formal creation of Nunavut in 1999. There were looks of amazement from the little kids when they heard the fine Inuktitut of Laurie Dexter as he addressed the crowd in the school gym. Many Inuit whalers made their homes here after the collapse of the whaling industry in the late nineteenth century, and many Inuit in Coral Harbor were obviously the product of mixed parentage. British and American sailors whaled in the eastern Arctic for one hundred years. Contact with crews of a whaling ship coincided with the epidemic-related deaths of the last of the Dorset people on Southampton Island one hundred years earlier.[13] This now is ancient history; we were treated with a certain disdain when we asked some kids how they got to their school's soccer and hockey matches: "by airplane of course." There were ATVs (four-wheel-drive all-terrain-vehicles) and a few trucks, but the roads led only around town, as in every Inuit hamlet. Travel is by plane, and by boat or ATV in summer, or skidoo in winter.

Skies were blazing orange in the evening as we crossed Hudson Bay. Our destination was Arviat (formerly Eskimo Point), a hamlet of about sixteen hundred on the western shore of Hudson Bay. Tourist ships are rare in Arviat, and we visited here only because the airport in Churchill, our intended point of departure, was closed for repairs for the summer. The people here are survivors or descendants of survivors of the great starvations endured by the Caribou Inuit, particularly the Ahiarmiut, and the Paallirmiut.

Before we left the ship, we heard from Elisappee, a surviving daughter of Kikik, whom Farley Mowat described in his three books.[14] This middle-aged school principal told her story in excellent English, bringing tears to us southern travelers. In 1958 Kikik's camp on North Henik Lake was decimated by starvation, and then by a murderous rampage of her deranged half-brother, who shot and killed her husband and attacked her. She was able to kill him with a knife, temporarily saving her family. Fleeing with her five small children, she attempted to walk at least fifty miles to a trading post at Padlei, carrying in her caribou-skin amautik her smallest child, and pulling two others on a sled. They were completely destitute, and had been forced to eat some of their skin clothing and tent to survive. It was February and bitter cold. They had no food and had not eaten for days. Exhaustion overcame her, and she was forced to abandon two of her children in a sealed igloo. She could not carry all of them, and unless she left some, all would perish. Later they were rescued by an RCMP officer, who headed back to search for the abandoned children, finding only one still alive. Kikik was tried by the white government for murder of her child, but was judged innocent by an understanding court. Soon after the Kikik tragedy, the government moved the survivors of those camps to Arviat. Kikik's youngest child was diagnosed with tuberculosis a few years later. She was sent south to a sanatorium, where she learned English, forgot Inuktitut, and was renamed Elisappee. She rejoined what was left of her family in Arviat, but did not learn of her mother's story until she read it in Mowat's books. Those who knew the story kept it private for decades. Elisappee was shocked on hearing her history, and was coming to terms still with all of the implications. Her family story was only part of a much larger story of suffering, now fortunately a thing of the past.

The land here is absolutely flat, only a few feet above the water of Hudson Bay, and with none of the great natural beauty that we saw on Baffin Island.

FIGURE 6 Elders of Arviat, dressed in their hand-sewn caribou-skin and beaded atigit.

There is beauty, but it resides in the people. It took us awhile to navigate in Zodiacs through the extensive reefs that extend for miles off-shore of Arviat.

Six elder women were standing together, waiting to greet us. They were a splendid sight attired in museum-quality caribou-skin *atigit* (inner parkas[15]) with beautifully colored and intricate beaded decorations, some with brass head pieces or colorful berets. We were told they wear these atigit only on special occasions. Each atigi took at least a year to make, and young women no longer were willing or able to do this work. The old women's faces were deep brown, and some were etched with wrinkles of perpetual sorrow despite the festive moment of our arrival. How many of their children had died because of starvation or various infectious diseases acquired from the white man?

Arts and crafts were for sale in the community center, a special opportunity for the Inuit to augment their government stipend, since visitors were rare. There were wall hangings (embroidered appliqués on a felt or stroud background) on the walls, and stone carvings on the floor as well as tables. We recognized a partially abstracted small piece with heads of a family group emerging from dull, rough, gray stone as one done by Lucy Tasseor. We exclaimed, "That is a Lucy," and almost immediately were joined by a short woman who spoke excellent English.

"Would you like to meet her? She is my mom."

Mary brought her short, sturdy mother over to greet us. Lucy speaks no English but solved the communication problem by reaching out and embracing Joyce warmly in a great hug. Lucy's hands were obviously arthritic, and we wondered how she created her carvings. Mary's excellent English was the result of being taken south to a tuberculosis sanatorium in her youth.

Someone told us Arviat is the friendliest hamlet in the North—not the most beautiful, but the most welcoming. When Mary asked Joyce to remain behind to stay with them, we were surprised and more than touched. We gathered on the shore, preparing to leave, when a man began drum dancing. Joyce asked if Mary could dance. The reply was, "Women can dance only if asked." Having been asked, Mary received the drum, and danced with skill and gusto. As we departed in Zodiacs, the old women in their marvelous beaded atigit lined up and waved. Beautiful in their bearing as well as their clothes, they were unforgettable.

Rankin Inlet (now renamed Kingiqsliniq) was the final destination, a relatively short distance by ship north of Arviat. A large former Cold War airbase, it handled big jets that could take us home. Several excellent wall hangings, as well as beautiful skin clothing that was illegal in the United States and totally impractical in North Carolina, were on display on the wall of the room where we gathered. We could not resist the opportunity to acquire an outstanding wall hanging.

"Irene will be so happy," said the very sweet Inuk saleslady, which made us feel happy, too. The Inuit are so easy to like; their warm, accepting, and patient manner left us hungry for more.

3

THE HIGH ARCTIC: RESOLUTE TO GREENLAND

Only the Air spirits know
What lies beyond the hills,
Yet I urge my team on
Drive on and on,
On and on!
INUIT SONG[1]

AUGUST 2002 A chance to cruise on the same ship through the high Canadian Arctic in August 2002 proved irresistible. The communities of the high Arctic may be of less interest in terms of Inuit art, but the land and waters are quite different, and much of the grand and mysterious history of exploration of the Arctic occurred here. Our youngest son, Mark, and his lovely wife, Alica, were free to travel, and they accompanied us.

A chartered flight from Ottawa to Nanisivik at the northern tip of Baffin Island was cancelled by severe winds that blew out the radar at the airport in Nanisivik. This hamlet offers a deepwater port; it was a former military airbase and the staging area for a now-defunct mining operation in the nearby community of Arctic Bay (Tununirusiq). Later we learned that severe winds are commonplace in Nanisivik, and many of the buildings are dome shaped and held down by cables to survive the winds. We waited one day in Ottawa and then two more in Iqaluit, the capital of Nunavut, on southern Baffin Island, for another charter flight to be arranged—this time to Resolute. The ship would be relocated near Resolute (Qausuittuq) to pick us up. We were on "northern time," when weather rules and one needs to be patient with often inevitable delays. Most of the people in the group were rather relaxed; what else could one do? One family freaked, the husband shouting with anger. The Inuit would not approve of such behavior. "*Ajurnarmat*" (there is nothing we can do about it), they would say. The rude man and his family departed.

The days of waiting provided time to see collections of Inuit art in the museums of Ottawa, and to get to know Iqaluit. Iqaluit is on Frobisher Bay, named after Martin Frobisher, who made three exploratory voyages to this place in the latter part of the sixteenth century.[2] In addition to engaging in lethal combat with some Inuit and taking hostages to exhibit back in London, he was fooled by the iron pyrite in the area, thinking it gold rather than the fool's gold it turned out to be. We marveled at the igloo-shaped church with its wonderful Inuit wall hangings and artifacts, all later destroyed when the church burned. Aaju Peter was to accompany us on this trip also, and we were delighted to be invited to tea in her home, a commodious modern apartment overlooking the harbor. Two of her children, her partner, and a neighbor who provides caribou for her winter larder joined us. Alica was glamorous when she tried on one of Aaju's handmade sealskin amautiit. Conversation drifted to the effects of the ban on sales of sealskin products to America and Europe, about the lack of understanding of the importance of sealing to the Inuit, and the damages Greenpeace had done to their economy.

Sylvia Grinnel Park near town was a grand place for a walk. There were wonderful views of unspoiled tundra, a beautiful river with a waterfall, and small Arctic plants already turning deep red in the early fall weather of August. Many Inuit families were camped in tents in small groups up the river, too far to visit. It did not escape us that they felt it necessary to create a park to protect this place from more development. In the more isolated hamlets, the entire countryside is one vast, open, unspoiled "park."

This growing capital of sixty-five hundred is a mixture of whites and Inuit, military and drunks, multiple hotels, government buildings, the Arctic College, and rapidly expanding clusters of homes built ever farther from town center. Iqaluit reminded us a little of suburban sprawl in the United States. In fairness, however, this is a city in which educated Inuit lawyers and politicians are emerging, where many of the political and thought leaders of Nunavut congregate. A taxi ride to "the most northern golf course in the world" revealed a few flags marking some old white rugs on the bare rocks and tundra. We recalled reading an article by a travel editor about his visit to Iqaluit, whose hopes for a romantic northern experience were dashed by reality. He did not go far enough north to find what he sought. It was a relief when we learned that a flight to Resolute had been arranged. Three days of the eleven-day trip were already gone.

Baffin Island looked entirely barren from the air as we headed northwest to Resolute, on Cornwallis Island. Our imaginations ran wild; it seemed forbidding below us. A nineteenth-century novel by Jules Verne[3] that Joyce found while waiting in Iqaluit describes the conditions in this area of the high North in imaginative detail. Numerous true-life adventures of explorations in this region[4] are even more exciting. Crossing the Baffin coast and heading over the waters of Foxe Basin, and then Lancaster Sound, north of Baffin, ice was densely packed as far as one could see to the horizon. Global warming may be here, but there was still a lot of ice. The entrance to the Northwest Passage stretched out directly below us; history came alive. We passed over entirely sterile looking brown land as we landed on an all-dirt strip at Resolute, at almost 75° north. Expeditions to the North Pole or to Ellesmere Island use Resolute as a convenient staging place. Almost nothing grows on this land, in a spot where summer temperatures are only briefly above freezing. Airplanes make this trip so easy that one forgets just how difficult and perilous it once was to get here. Less than two hundred years ago, many brave men suffered and some died trying to find a way through to the west. Many ships were trapped in the ice, sometimes for several years.

Resolute was uninhabited until the airbase was constructed after World War II. Some Inuit families were relocated to a spot a few miles distant in 1953. One reason for their relocation was to help establish Canadian sovereignty over the Arctic archipelago, as Norwegian, Russian, and U.S. interests threatened to compete for claims to mineral and oil rights in this area. The Inuit had little experience with hunting animals of the sea, and life was very hard: the cold and darkness were very different from the low Arctic community of Inukjuak (formerly Port Harrison) in Nunavik from which most came. Particularly telling was their name for this place: Qausuittuq, meaning "the place with no dawn."[5] It is hard to read about their treatment, which included forced separation of families.[6] The Inuit were promised they could return to their original communities, but these promises were never kept.[7] They were victims of a bureaucracy that understood them poorly and used them for its own purposes, and which many years later attempted atonement by paying reparations to survivors.

Our leaders were in a hurry to continue, so we did not visit the Inuit community, located only a few miles away. We traveled by old school buses to Zodiacs at the water's edge, passing big satellite dishes aimed low on the horizon, scattered *qamutiik* (wooden sleds, pulled either by dogs or motorized skidoos), and debris. The gleaming white ship was anchored many hundreds of yards offshore, separated from us by thick fields of drifting pack ice. Donned in life jackets, wool caps, gloves, and parkas, we twisted through the ice in Zodiacs, grateful that the tides had not packed the ice more densely. The day was clear, and the skies were an intense deep blue over the dry, barren browns of the hills—a high northern desert. An icebreaker was helping a freighter get through the notorious ice of Lancaster Sound. We followed it through the broad ice fields, which turned gold in the low light of late evening. A polar bear was seen on the ice, much too distant for photos. The sun drifted lower until it disappeared entirely, leaving us in gray gloom, with just a hint of light at the far horizon under the clouds. It snowed briefly.

The light was dim when we approached Beechey Island about one in the morning. We stopped at this late hour because we were so far behind schedule. In retrospect it was our great fortune to visit under these conditions. Many have stopped here, but few have disembarked at this hour. The low light contributed to an otherworldly feeling. Ahead of us black cliffs rose, rising vertically for hundreds of feet with slopes of scree at their bottoms. Slim bits of fog sailed across the cliff faces. The entrance to the harbor was guarded by cliffs on two sides, and the seas within the drifting pack ice were entirely calm.

Beechey Island is one of the most famous places in the Arctic. John Franklin and his crew of 129 men on the ships *Terror* and *Erebus* made their first winter camp here in 1846–1847. They were searching for the Northwest Passage, and the quest ultimately ended in all of their deaths. The first three crewmen to die were buried here. An old engraving of these graves, which we had found on Cape Cod, provided our own documentation of them. Franklin's remaining crew members sawed out of the ice the next summer, but were frozen in again by September 1847 somewhere near King William Land, west and south of this site. Their most severe suffering started then, including starvation. They were never seen again by Western eyes, but later evidence showed they had tried to march out carrying literally tons of stuff, including silverware and fancy dress uniforms. Before it was over some of them resorted to cannibalism. For years, other ships searched for them,[8] but it was Inuit who eventually found the first evidence of what happened. Inuit oral history passed down for generations stated that they appeared scary, were very thin, and wore very poor clothing. Bodies were found that appeared to have been mutilated; starving men were seen carrying body parts.[9]

The beach stones under our feet were almost entirely barren of life, except in small, round patches near old cans left by the Franklin party. The cans once held pemmican, and eventually rotted, releasing their nitrogen into the barren soil and enabling ring growth of tiny flowers almost 160 years

FIGURE 7 Empty, rusted tin cans on Beechey Island, where Sir John Franklin and crew established their first winter camp in 1846. Released pemmican fertilizes the bare tundra 160 years later.

later. Things decompose slowly here. We could see the solder lines on old rusted cans, now believed to have caused lead poisoning that contributed to their deaths, attested by postmortem samples taken from these three graves. Those cans with their crude, heavy solder lines were testimony to the problems caused by a policy of low bid wins; a new contractor had been used to make these fatal cans.[10] The ruins of Franklin's long wooden house were still visible on shore. We were chilled as we looked out at our ship, its lights glowing, and warm inside; we tried to imagine how very cold and desolate it was during the winter months of twenty-four-hour darkness in Franklin's time. It was a ghostly sort of place. We finally got to bed about 4:30 a.m.

Lancaster Sound is broad, long, straight, and deep—a nautical superhighway in an otherwise highly congested archipelago. It was discovered and named in 1616 by William Baffin, who also correctly outlined the geography of the entire region north of Davis Strait; Baffin Bay and Baffin Island are named for him. The English left much in the North, including place names for everything. Two hundred years later John Ross entered Lancaster Sound on a voyage of discovery, and suffered the unfortunate mistake of seeing a mirage of mountains,[11] turning back because he thought there was no way through; only one year later other explorers proved that the passage west was open, and indeed was the gateway to the long-sought northern passage to the Orient.[8]

We marveled at the high cliffs and glaciers along the Devon coast. The day was beautiful, with bright sun etching the snow and fog along the cliffs. Mark pranced about on deck like a young deer, unable to control his enthusiasm. We pretended that his extra energy came not from youth but from sleeping through the stop at Beechey Island.

Our immediate destination was Dundas Harbor on Devon Island, the site of another brief and failed experiment in Inuit relocation in the 1930s, and of intermittent RCMP posts, both designed to help control the Canadian rights to these lands. Devon Island is the largest uninhabited island in the world—too cold here to sustain life. The Dorset and Thule lived on Devon successfully, apparently when the climate was slightly warmer. The fog was dense and delayed our entrance to the harbor, near midnight once again. The twenty-four hours of bright summer daylight were now morphing in mid-August into evenings of deep dusk with just enough light to get around easily. The ship was barely visible from shore, hidden in fog, and looking small against the high cliffs rising above the horizon. We hiked along the tundra, admiring the vistas, wrapped in somber blue-gray Arctic light, fascinated by the colorful rocks and lichens. Piles of dried old musk-ox and caribou dung were scattered about. Mushrooms and a few grasses were among the few living things seen. A lonely and very cold-appearing walrus was waiting for something by the shore; he looked back our way, quite aware of us, but held still until our slow approach finally forced him back into the water. Old Dorset and Thule homes were found, overlooking the ice in the bay. Not much needed to be said. Dawn came soon, rising above the black hills. The light was a beautiful backdrop to Aaju, standing silently with a rifle slung over her shoulder, always on the lookout for polar bears, illegal to shoot except for protection.

The skies were a magnificent panorama of changing pastels, first a creamy gold, then rose pink, and later blue-gray with mounds of low clouds casting their shadow on calm waters as we cruised south into Navy Board Inlet, on the way to Eclipse Sound and Pond Inlet. A large glacier and perpetual snow caps marked the black contours of mountainous Bylot Island, which forms the northern boundary of Eclipse Sound. Years ago, Peter Freuchen and a small band were trapped on a bit of ice that broke off from the main pack, and they drifted all the way up into Lancaster Sound before they were able to get back to land.[12] Baffin Island lies to the south, flatter than Bylot Island. There are low mountains, however, behind the hamlet of Pond Inlet (Mittimatalik), a very picturesque Inuit community and another gateway to the far North. Traveling in this manner is exhausting, since the very best colors at this time of year are seen at the late evening and early morning hours, and our best stops were made at midnight. We were delightfully sleep-deprived.

The transit through Navy Board Inlet lasted only hours, but our memories lingered there, recalling the wonderful book *North into the Night* by Alvah Simon.[13] In 1994—he and his wife sailed a thirty-six-foot sailboat into this same spot, and purposely overwintered on the frozen shores of Bylot Island in a sheltered cove named Tay Bay. His wife had to leave during the depths of winter to attend her ill father, and Simon was alone through the three months or more of complete darkness, in temperatures as low as −50°F (−46°C). He was visited by Inuit friends and helpers whom he had met in Pond Inlet, some one hundred miles away over the ice by dog team. He survived the ravages of the ice on his fragile boat and was gratified by the acceptance of the Inuit, who understood the difficulties of being alone so long in such conditions. Through his words, one can almost experience the wonders of the return of light, and the beauty of seeing, in such an intimate way, the lemmings, foxes, bears, gyrfalcons, ducks, loons, snow geese, ravens, and other life as warmth returned to the frozen land. Simon had great respect and affection for the Inuit. Their stoic strength and bravery impressed him, as did their ethic of community first, with self second. Some of them were beginning to renounce Western ways for a return to life on the land, depending instead on hunting for narwhals and caribou, fishing with nets for char, harvesting bird eggs, and living the old way. We envied Simon's experience, but doubted, even at our youthful age, that we would try it.

Bylot Island is full of birds and other small animals, but the last caribou were killed many years ago by Inuit hunters after they acquired rifles.[14] The big animals are in the water; there is a chance to see narwhals near the floe edge when they migrate through in late spring and summer. Huge bird rookeries are found on its cliffs in the spring, which are of particular interest now that science has found that seabird guano is highly concentrated in organic and inorganic toxins that originate in the industrial South, resulting in great accumulations near their breeding sites.[15] Such toxins find their way into the Arctic food chain and become concentrated at the top of the chain in predators such as the polar bear, seals, and the top predator of all, humans.[16] The result is a worrying paradox: unusually high levels of toxic organics in nursing Inuit mothers and in bear populations in this, supposedly, the most pristine place on earth.[17] Indeed, we are one planet, and nothing is safe from the byproducts of southern industry. This is something worthy of our thought and hopefully our actions, but the problems run very deep. And there are other sorts of toxins up here, including plutonium bombs that were lost after a U.S. bomber crashed in deep water off northern Greenland, not too far north and east of here across Baffin Bay.[17]

Pond Inlet was framed by snowcapped peaks rising high above town, and by an iceberg stuck in the harbor. The hamlet sits on a rise looking out directly at Bylot Island across Eclipse Sound, and seemed almost close enough for a canoe paddle to Bylot, although in the very clear air and without any visual guides it is very easy to underestimate actual distances. One can see a very long way here in the clear air that is almost completely free of the smog that we now take for granted in most of the more developed places to the south.

It was from Pond Inlet that the shaman Qidlaq and a team of trusting followers began in 1862 their amazing trek north past Bylot Island, across

Lancaster Sound, Devon and Ellesmere islands, and Smith Sound to reach the shores of Greenland, fulfilling a shamanic vision of his that they would find other Inuit peoples there.[18] Some of them lost faith and turned back, but many stayed with him throughout the incredible journey. The entire round trip took seven years, and Qidlaq and several others died on the way back. The people they found in Greenland had similar language but had lost knowledge of use of kayaks and sinew-strengthened bows, so long had they been separated from other people. Many Inuit in isolated upper western Greenland are descendants of these brave travelers.

At the community center of this modern version of Pond Inlet, no shamans were practicing, but we saw throat singing, a performance of Arctic games, and had our picture taken with a short, rotund Inuit woman in a beautiful amautik whose beaming smile suggested her pleasure in showing off her hamlet. Mark found a room full of computers that had been donated by Microsoft, his employer, and he emailed Bill Gates to say how important this gift was to the community. To his surprise he quickly received an email that, although unsigned, seemed to be from Gates based on the terse style, said Mark. He was more than a little impressed with himself, and so were we. A walk along the shore took us past many smiling, attractive kids, and homes with caribou antlers on the roof, to the Salmon River and an old Dorset site. A stocky Inuit couple stood on shore next to orange gas cans looking at their small open boat, evidently preparing to take a trip of their own. Our much larger ship was anchored in the background, and we wondered what they thought of us. We passed a grave that held the remains of a white man who had come here many years ago, and who was murdered by decision of the community because he was such a disruptive influence. We tried to behave. Before we left, we entered a very small Anglican church, apparently the church of our former trip leader Laurie Dexter, who was a minister here for a decade. This is a very isolated and perhaps lonely place to live, but it is certainly one of the most beautiful.

Narwhals were not in evidence, but we did not have time to look for them properly, and may have come too late. They are abundant in the waters around northern Baffin Island and Lancaster Sound, especially when the ice is breaking up. These waters are said to be particularly thick with life, including polar bears, seals, polar cod, narwhals, and many species of birds, in a land that to the unknowing looks too harsh and extreme to support such life. Reasons for this richness of life are unclear, but possibly have to do with runoff from Bylot and Devon islands, and upwelling of deep waters that provides nutrients for the small and big animals.[19] Life depends in great measure here on the ice, which is coated on its underside by algae, which feeds clouds of zooplankton, which in turn feed the cod and other fish that sustain the larger animals and birds that feed on them. Loss of the ice would devastate this rich ecosystem.

As we crossed iceberg alley on the way to Greenland, fabulous bergs appeared, one of which we circled to fully appreciate its great open arch. There were many happy times with much laughter among the four of us, and many moments of shared mirth with Aaju and the other Inuit woman on this trip, Meeka Kilabuk. Meeka until recently was a member of the new Nunavut Parliament at Iqaluit, and it was said she was a real force, outspoken and opinionated. She told Joyce that she was impressed with her and Mark's interaction, a statement about the importance of family to the Inuit. Mark was a big hit as the male model for caribou skin clothing made by Aaju, displayed for the benefit of the ship's complement in the dining room. The weather was very calm, and we experienced none of the horrendous storms about which we heard much. A few sei and fin whales and a lone breaching humpback enlivened the scene.

The shore of Greenland was stunning—high, dark rock topped by white ice cap, with bergs of all sizes and shapes in the water and little white tents on shore, housing people living the summer months on the land. The Island community of Illorsuit at the entrance to Uummannaq fjord was where Rockwell Kent spent many of his years, painting and making wood-block prints.[18] The community of Uummannaq was set on an island with bare rocks and colorful little houses, framed against a large steep hill. Another small island was close, separated by a bay filled with icebergs. The homes were painted in colorful reds, blues, browns, and greens, and were tied to the rock by cables. An old sod house was preserved, as was a sealskin umiaq sitting on a wooden frame outside under a protective roof. Inside in a small museum there was a well preserved and beautiful sealskin kayak. This was one of the principal places where Gretel Ehrlich lived and from which she wrote beautifully phrased books.[18] Most of her time here was in the depths of winter, with all the cold and darkness, ice and snow. Only by reading her glowing sentences do we get a sense of what it is like to be here in winter. She reported that was her best time, a time when the darkness let her imagination run free.

Many Inuit were shopping for beluga whale meat, and slabs of whale with their white skin were going fast at a market in Ilulissat. Mark treated us to an old Russian helicopter ride up to the head of the ice fjord to see the great glacier coming off the ice cap. Here we nestled among the rocks, looked upon endless stretches of white glacial spires, and basked in the sun at temperatures just above freezing. Near the edge of the glacier, small meltwater ponds were rimmed by heath plants now turned bright red as fall approached, and Arctic cotton blew in the breeze. The surface of the glacier was jagged and covered in a gray dust, and it was a very long way across the breadth of the frozen ice river below the face of the glacier. All was quiet.

Out in Disko Bay we could see gigantic newly launched bergs still topped by jagged gray dirty ice before they melted, rolled, and cleansed themselves of the accumulated grime. The air was bright on a sunny day, and the visual and emotional impact for us was quite different than in the mysterious low light and fog of our previous visit. Perhaps the first time is always the best. Or maybe the light conditions are the most important thing. Gulls sat on bergs, white against blue. Later some of us enjoyed single-malt scotch with a chunk of crystal-clear ten-thousand-year-old ice taken from the base of a small berg.

Back again at Kangerlussuaq, we were driven out of town where we saw musk-oxen in the far distance. The tundra was colored a variety of reds and yellows in the full glory of autumn, not long before the onset of the very long winter. We found a place where the four of us could sit quietly, on the rocks overlooking a long valley. We were at peace. Eventually Mark and Alica left their photographs as a screen saver on our home computer, so we now see the two of them flashing across the monitor, cavorting, acting silly, carefree, and happy against the backdrop of some of the most beautiful places one can imagine.

Two cruises, first through Hudson Strait and across Hudson Bay, and now through the high Arctic gave us a bird's-eye view of the geography of Nunavut. Our fellow passengers were enjoyable, and the vistas and photographic opportunities bordered on the incredible. Our impatience grew, however, with the superficiality of many of the experiences. It was time to get off the cruise ships, to meet and learn more about the Inuit.[20] The art was influenced by the land, but it was Inuit who made it.

FIGURE 8 An early unsigned and untitled wall hanging ca 1973 from Baker Lake attributed to Marion Tuu'luq. It may depict a vanished way of life in igloos, but its meaning is enigmatic.

4

WHERE THE BULL CARIBOU GATHER: PANGNIRTUNG

*If you ever feel the vast space of wilderness, you will see what
I am trying to say in words. Words are not enough.*

LAZALUSIE ISHULUTAK[1]

JUNE 2002 Standing on a barren hillside hundreds of feet above the hamlet, we could see the frozen fjord flanked by high, snow-covered mountain peaks. Wisps of fog drifted over the ice, partially obscuring the blues of the mountains. The sun was bright but low on the horizon. A solitary inuksuk stood guard on the hillside, an unambiguous declaration that this was an Inuit community.[2] This was Pangnirtung, simply called Pang, one of the most beautiful places in the Arctic. The Inuit call this former hunting camp Panniqtuug, meaning the place where the bull caribou gather. It is a community of perhaps twelve hundred people, almost all Inuit, nestled off the north coast of Cumberland Sound on Baffin Island, just below the Arctic Circle. We were here for four days.

We tried to enter by cruise ship the year before, but ice frustrated the effort. When our friend H. G. Jones told us he was going to Pang to help celebrate the birthday of his old friend Taukie Maniapik, and invited us as fellow Inuit enthusiasts to accompany him, we jumped at the opportunity. H.G. had been here perhaps thirty times, and knew the people well.

H. G. has been our guide, and we owe much to him. He was born near Kill Quick, North Carolina, and grew up in a home without books during the Great Depression. That modest beginning did not suggest his future accomplishments: the Humanities Society 2002 North Carolina Award recognized his lifetime's work as archivist for the Department of Archives and History, and curator of the North Carolina Collection at the University of North Carolina at Chapel Hill. In 1971 he was unsure what to do with a rare personal vacation, and ended up traveling to the North, leading to a lifetime passion for the Canadian Arctic and the Inuit. He is an expert on the history of the Inuit and has accumulated a large collection of Inuit art, much of which he bought directly from or was given by the artists during his travels. Looking at his collection brought forth many stories of the North and of the people whom he has met, particularly Pangnirtung's Andrew Qappik. He befriended Andrew as a teenager and collected every one of his prints over the ensuing years. All of those prints now are in the collection of the Winnipeg Art Gallery (WAG), a gift from H. G.

The three of us left in the steamy heat of a North Carolina June to fly to Pangnirtung, via Ottawa and Iqaluit—a trip of two days. The final leg was a mere two hundred miles, and our fellow travelers all were Inuit. Most smoked, a habit acquired many years ago from white visitors. The bare rocks of the high plateau of southern Baffin Island were dusted with snow, and etched heavily in parallel grooves by the glaciers that once covered the land. Cumberland Sound appeared, and then a beautiful fjord solidly covered by ice. The plane settled easily on a gravel landing strip essentially in the center of town. The small airport building was decorated inside with a beautiful large tapestry from the weave studio, and outside by a wooden plaque of a massive bull caribou made from

FIGURE 9 A solitary inuksuk overlooking frozen Pangnirtung Fjord.

one of Andrew Qappik's designs. Along the distant shores in every direction, there was no sign of human habitation, but in Pang there was a lot of happy human activity, as Inuit families and friends greeted arriving passengers. Little kids ran around, smiling brown faces peered from amautiit; a swarm of people surrounded us. Their excitement at meeting family and friends was electric, a kind of force field that energized us too. People departed swiftly in old trucks or on ATVs, we among them on our way to Auyuittuq Lodge.

After hearing tales from H.G. about honey buckets and reading historical treatises on Arctic human waste disposal, we wondered what facilities would be available for us. A private room with toilet was gratefully accepted. Generous-sized meals included local char. The Inuit women who prepared and served these meals seemed pleased with their jobs and anxious to communicate with us. An eclectic small number of white men were lodgemates. Among them was a retired Canadian family doctor who was returning to Pang for a vacation. He had vast experience doctoring in the North, and clearly loved the Inuit. One reason he so admired the Inuit was the directness of their gaze when they came to see him. His stories enriched many mealtimes.

Walking through town with H. G., we were startled by the sudden appearance of a corpulent figure with a beaming broad face walking along beside us. H.G. turned and with great joy greeted his unofficially adopted Inuit son, Andrew Qappik. Though shy and reserved, their joy at seeing each other again was obvious. After some friendly banter, Andrew continued on his way to the Craft Center. We ambled over to the cemetery, perched on the edge of the ice-covered fjord, and looked at many small gravestones in the chill breeze. Many of H. G.'s old friends rested here. Nearby, a few dwarf Arctic willows were in bloom, their catkins backlit by the low light, their branches spread out only inches above the ground, a means to escape the wind. By staying close to the ground they also were able to take advantage of the zone of summer warmth just above the ground. Nothing else was in bloom. Survival here is a challenge.

Row houses lined the sides of the two streets that run through town. They were built by the government from prefabricated materials brought here by ship in the summer thaws, and were positioned on stilts to prevent the heat from melting the permafrost. Many were held down by cables that passed over their roofs, to prevent fearsome winter winds from literally blowing them into the fjord below. One building jutted halfway over the ice, witness to a storm that unhinged and almost destroyed it. Imagine winter winds of eighty to one hundred miles an hour and temperatures of $-40°$ F. We marveled at the ability of the Inuit of the past to survive in igloos and skin clothes in the outlying camps. Even now it is dangerous to wander outside in a blizzard; the snow was so blinding that someone froze to death within yards of home only a year or two ago.

Large char hung outside homes to dry, the flesh red and the skins silver. Ringed sealskins were drying on racks at the sides of some homes. skidoos and qamutiik nestled close by. Boats were scattered about near the shore edge. The bottom of one of them was covered in fresh blood, probably from a recent seal hunting trip to the floe edge.

Across from the lodge close to the shore were several red-roofed white buildings bearing the sign "Hudson's Bay Company Old Blubber Station";

underneath what we presumed was the same designation in the syllabic text of Inuktitut. The use of syllabics is over one hundred years old, and was rapidly adopted after a missionary introduced it, first to the Indians and then the Inuit. Syllabics is still widely used, especially by the older Inuit and in the more isolated communities. Lying close by were two wooden dories on their sides, one in an advanced state of decomposition. This was a great whaling place in the nineteenth century, a haven for bowhead whales (Greenland, or right whales). Now the former blubber station is empty.

Understanding Pangnirtung is impossible without understanding the influence of the whalers.[3] The anthropologist Franz Boas lived and studied in Cumberland Sound in the 1880s. He estimated that in 1840, when the whalers first arrived in force, there were about eight main hunting communities scattered around Cumberland Sound, each with about 200 people, for a total estimated population of about 1,600 people. Only seventeen years later a Moravian missionary estimated the population at only 300, and in 1883 Boas counted 328 people.[4] He attributed the decline in population to the introduction of infectious diseases such as diphtheria, polio, measles, influenza, tuberculosis, smallpox, syphilis, and others, a very familiar tale after the first contact of indigenous peoples with white whalers. The Inuit eagerly helped the whalers, both in hunting and in maintaining the primary whaling communities on Kekerton Island, and Blacklead Island near the head of Cumberland Sound. Eventually the great whales were all but exterminated, the whalers left, and in 1921 the first HBC trading post was established along with the hamlet of Pangnirtung. Shortly thereafter the area hospital was built in Pang. The Inuit persisted in a more or less traditional life. After the dogs of the Cumberland Sound area were devastated by an outbreak of distemper in 1962, and difficult ice conditions caused periodic starvation, many Inuit gave up and moved from their camps into Pang.

The whalers used to summon their Inuit helpers from their camps with a cannon shot. A cannon remained in the middle of the hamlet, a reminder of the whalers and their influence. British and American whalers changed the culture, and left behind not only memories but also some of their genes. The only whaling now is for food and ivory, conducted by individuals in small motorboats. Only belugas and narwhals are hunted, and narwhal hunting is limited by quotas. We witnessed a proud hunter bring a six-foot narwhal ivory tusk to the craft shop, where the manager purchased it for six hundred dollars. A few bowhead whales still gather here in summer, but are protected by a limited annual hunt for all of Nunavut.

Many boats, skidoos, and qamutiik sat on the ice in the fjord below the lodge, ready for trips to the floe edge. A strong, thick man carried four fifty-pound gas cans at a time, two in each hand, over the huge boulders and the large rough and irregular chunks of shelf ice at the edge of the shore. It was hard for us to navigate the same ragged ice shelf with nothing in hand but a camera. Periodically the hunters departed in their skidoos, as likely in the gray midnight light of June as the bright light of midday. We knew when they left because of the roar of the skidoos. We also could hear the laughter of little kids out on the streets late at night, taking full advantage of the return of the sun.

The June ice was beginning to rot. There were pools of meltwater sitting on the ice of the fjord, and one wondered how the hunters knew when the ice was too soft or thin to support the skidoos. Although they can read the ice conditions well, many Inuit have died after falling through the ice, which looked fragile and fragmented to us.

The light was beautiful, especially late in the evening when the sun's low circling path passed behind gaps in the low mountains on the opposite shore, casting a focused beam of bright light onto the ice, illuminating wisps of fog hovering overhead, all fronting the gray-blue mountains. The pools of water on the ice shone brightly. One afternoon a midday fog partially obscured the sun, showing the moon on its shoulder, recalling the Inuit myth of the birth of the sun and the moon.[5] The changing light and fog are a regular part of the Arctic, beautiful at times, but also extremely hazardous to travelers.

Wandering around town, we visited all the stores, which was easy as there were very few. The Northern store was the main place for shopping, and there were two co-ops as well. Curiously, the Northern stores now include all of the former Hudson's Bay posts throughout the Arctic, a final victory in a centuries-old clash between the two companies for commercial dominance in the North. Everything was expensive, including the outboard motors and skidoos. Signs in the entrance educated us about the dangers of polar bears, and how to practice safe sex. A community center for elders hosted an exhibit of cultural artifacts.

The health center looked modern and clean. Resident nurses supplemented with occasional one- to two-week visits by a southern physician and a dentist administered care. Patients with serious illnesses were flown to a hospital in Iqaluit. We recalled reading about the history of Pang's Anglican hospital, now closed, that for several decades served the people of the eastern Arctic. Otto Schaeffer practiced medicine and operated as best as he could under the conditions in this hospital in the 1950s, while we were at university, and most of the Inuit still lived in camps. He became a famous figure, as did his very able and strikingly handsome Inuit teacher and helper, Etuangat. On their long trips to the outlying camps, Etuangat and Schaeffer became fast friends, facilitating medical practice here that was literally an adventure.[6] A long and dangerous trip by dog team over the ice in the dark winter cold was required to care for a pregnant mother suffering a difficult birth in her igloo far from the hamlet. When surgery was required in the little mission hospital, there were no consultants to help if things went badly.

Chartreuse posters all over town in both English and syllabic Inuktitut announced the forthcoming celebration of Taukie Maniapik's ninety-sixth birthday celebration. Taukie's youngest daughter, Lucy, came home from Iqaluit for the occasion; she organized the entire celebration. One day Lucy visited us in the lodge, bearing very red and quite tasty freshly cooked seal meat, our first experience of this Inuit staple. Later we went with her to Taukie's home, which had many photos on the walls of H. G. and Taukie together over many years. Taukie was alert and conversant, but confined to a chair. Curiously, he spoke no English and H. G. spoke no Inuktitut, but they were great friends. How they managed this was a mystery, but their huge smiles on seeing each other told it all. Many family members were there, including a son who was mayor of the hamlet. A son-in-law who was an Anglican minister in Kimmirut recalled seeing us in town when we visited Kimmirut the previous year. A very satisfying caribou stew was served with tea. Everyone seemed comfortable with our presence, no doubt because we were with H. G.

Lucy surprised us with a comment that she was planning a vacation to the Caribbean with her partner, with whom she ran a successful computer shop in Iqaluit.

"Why not?" said she.

"I don't like to be cold."

Indeed, why not? We don't like to be cold either. Lucy can tolerate the cold, though; her partner told of their nonstop skidoo trip in January from Pang to Iqaluit and back, some two hundred miles each way. He allowed, in a frankly admiring tone, that "she is strong and tough."

The celebration of Taukie's birthday took place in the community center, crowded with perhaps 250 people of all ages. Introductions from dignitaries started things off, and there were gifts for Taukie and others, including us, to our amazement. H. G. was the star, as he showed an extensive set of slides that often brought gasps and applause when someone was recognized in old pictures taken decades ago. The audience contained many interesting people, including a very large young woman in training to become a member of the RCMP (she appeared up to the task); many young mothers with little kids in their amautiit; several old ladies in lovely sealskin *kamiks* (boots); a handsome bronze-skinned and athletic looking man in a truly marvelous and well-fitted sealskin coat; and Elisappee Ishulutak, now very old but still beautiful of face, bent over a walker and looking to be in pain as she moved. She once was one of the forces here in art, both in the print shop and in designs for tapestries from the tapestry studio. A chocolate cake inscribed in syllabic Inuktitut was served, and then the games began. What fun! One game consisted of a roll of the dice which signaled one of the ladies to leap into a circle and accumulate as many bracelets on her wrist as possible before another lucky lady jumped in. The winner was an old woman who was so supple that she could almost put her wrists on the floor, bending at the waist with straight legs. This is the position for flensing seals, and their active lives keep them flexible in ways that are not possible for persons who spend a lifetime working at a desk. There were gambling games and candy throws for the little ones.

The main game was a competition between half a dozen pairs of people; a woman serving a baby bottle to her male partner who kneeled at her feet. At the urging of someone, Fred got into the circle with Jevua as his partner. She laughed and urged him on as he frantically sucked at the bottle, which had an excruciatingly small hole in the nipple. The laughter was loud. Fred's cheeks grew sore, but with a well-documented type A personality he was determined to win, and a win it was. The prizes were a pair of work gloves and a heavy wrench. We finally convinced Jevua to take them home. "Are you really sure?" This was not an ancient game, given its use of baby bottles with rubber nipples, nor was it unique to Pangnirtung. Duncan Pryde described a similar contest in another hamlet forty years earlier:

> We'll let the women pick out the strongest men.... The one who finishes first in this contest will be the strongest. They all watched intently as I opened the box and handed each a baby bottle full of milk, nipples all prepared. I gave a bottle to each man. The sight of those three strong, tough-looking Eskimos sucking away on their bottles brought gales of laughter from the crowd. Irvana shouted: Look at that man, I wouldn't mind having him at my breast![7]

Jevua did not make such an offer to Fred.

Pangnirtung is known for its print shop and weave studio, centers of creative work for decades. Now housed in a contemporary new building that replaced the burned one, the Pang artists are poised for new demands from the southern market. Weaving was introduced in 1971, and tapestries remain unique to Pangnirtung among all the Inuit communities. Large weavings based on Malaya Akulukjuk's landscape drawings greeted us as we entered. Malaya was responsible for many of the early weavings of the spirit world, depictions stemming from her prior role as a shaman. The colors and textures of the tapestries are warm; their subject matter includes life on the land in all seasons, as well as fantastic images from myths.[8] Five women were working on the largest tapestry they had ever attempted, a ten-foot-by-twenty-two-foot piece designed by Joel Maniapik, which had to be finished for the Queen of England's visit to the parliament building in Iqaluit five months hence. Most tapestries were large and demanded considerable wall space. We did take a small piece home, designed by Andrew Qappik and woven by Kawtysee Kakee, whom we met. Although Kawtysee is deaf and mute, she has achieved style and skill in her weaving.

FIGURE 10 *Untitled*, 1971, Pangnirtung. Malaya Akulukjuk's design and Meeka Akpaliuk's weaving.

There was a sense of great pride among the weavers, and unambiguous talent. We came back later for a second visit. The women radiated a sort of warmth and cohesiveness that drew us in.

Church is important in Pang. Shamanic religious beliefs that once dominated their lives with myriad taboos were outlawed by missionaries and were replaced by devotion to either Catholic or Anglican Christianity, or in more recent times by other fundamentalist evangelical churches. The Inuit were quick to convert in most instances, trading fear of the unknown and of the various spirits that ruled their lives for belief in a more benevolent God. The power of saints replaced the power of amulets. We attended services at the Anglican church; it was moving to hear the hymns of our youth sung in Inuktitut with the fervor of ardent belief. A tall white priest whose speech sounded Scottish was accompanied by an Inuk priest, both in robes. They welcomed us warmly in English. The church was nearly full. Little kids ran around freely and were only brought under parental control if they became disruptive. It was a happy place, full of good cheer.

At a wedding in the same church, a beaming bride walked down the aisle in her white veil and blue gown that was much like those worn at the senior prom fifty years ago. Rice and confetti greeted the couple as they left church and were taken away in a car decorated with streamers and balloons and a "Just Married" sign to a destination unknown but certainly not far, given the limits of the road.

Inuit came to visit in the lodge, or to sell crafts or carvings. Visitors included Andrew Qappik and his wife, Annie, with whom we had a long talk. Some young women came to visit out of curiosity, and stayed for quite a while, talking seriously one minute and the next erupting into nervous giggles and laughter. One young woman seemed troubled and lingered to speak with us. Later we saw her in town with her baby, and she still seemed wounded. We learned from the Co-op manager that she was in trouble with the law again, for beating her husband after a drinking bout. After many sad and unhappy experiences with the ravages of alcohol, the community passed strict laws against importing alcohol.

Discussions in the lodge came around to the epidemic of adolescent suicides in the past year, usually by hanging from a bridge over the river just outside of town. The epidemic was stemmed by sending many of the youths out of town to live among people who were not trapped in a cycle of death and depression. Apparently, there was not enough to do, no hope for the future, lack of guidance from the former strictures on behavior, too many drugs, and too much glue sniffing: a very sad and worrisome problem. Similar problems exist everywhere, including our own community, yet there was a high level of concern here. The suicides were fresh in the minds of many people.

If this is the place of the bull caribou, where are the caribou? Inuit told us they disappeared for a long time, but now are appearing on the opposite shore of the fjord. The meat is important in a place where government checks are barely enough to survive. There are lots of char, which made us think about coming back after the ice leaves to fish for them at a camp down the coast. We asked about the limit on the daily catch. An Inuk companion did not understand, and finally scoffed at the question. Catch limits?

Fred dropped by the craft shop for a last look before we left for the airport for our return flight. As he turned to head back to the lodge to meet H. G. and Joyce, a man appeared suddenly from the back of the shop, waving his arms and indicating that he knew we were about to fly. It was the hunter who was so glamorous in the beautiful sealskin coat at Taukie's birthday celebration in the community center, Lazalusie, a son of the elder artist Elisappee Ishulutak. He spoke no English, but wanted Fred to wait so he could give him something. A few minutes later he reappeared with a copy of a long article written about him in the *Kitchener Waterloo Record*, October 1, 1987.[1] "Once I lived with the rhythm of the land. Life has changed so much.... Now we live with the ways of the south.... I know my life was happier and freer." Lazalusie lived on the land until about 1984, when he and his wife and seven children were among the last to move into town from the land. "I hunted caribou. You had to be perfectly healthy—everything perfect in your body. Hunting on foot, going out over the land for days, carrying the caribou on your back, you cry the rest of the season. But you get a supply of food and the skins give you warmth that lets you survive any weather.... To be a man meant the whole camp survived.... I was taught only to survive on the land. It is in me, in my roots.... I can't stop it. I can't change." Now, however, Greenpeace and the animal rights movement have resulted in injunctions against selling sealskins to markets in Europe and the United States, which has crippled his ability to support his family. He misses the open tundra, and being on the ice and sea. Fred was deeply touched by Lazalusie's wish that we should know this about him, and about the plight of most Inuit.

The flight out took us over the floe edge, where we saw a large circular pan that had broken from the main ice, floating free. There was a well-defined pair of skidoo tracks on the pan, going from one side to the other, easily seen even though we now were high above the water. No doubt the hunters crossed before the pan broke free, but we wondered. We recalled some of those who were trapped on broken floe ice, including Dr. Wilfred Grenfell off the coast of Newfoundland. He survived only by wrapping himself in the warm skins of his huskies, which he was forced to kill.[9] Even more dramatic was the experience of some members of the Hall expedition of 1871–1872. They were separated from the Polaris off northwestern Greenland, which subsequently sank, and found themselves on a four-mile ice pan. Drifting south for 193 days, they were rescued off Labrador after traveling over thirteen hundred miles on an ever-shrinking ice pan. They survived because Inuit with them hunted seals; a baby was born on the ice.[10]

Flying home, our heads were full of Kawtysee Kakee and her successful artistic efforts against many odds and Lazalusie in his beautiful sealskin coat, made of seals he caught and which his wife prepared and sewed. Both were cast adrift in a sense. For Kawtysee, weaving in the hamlet gave her a meaningful occupation, but Lazalusie felt marooned in the same hamlet. We also felt trapped in certain ways, and longed to get out onto the land.

5

AN UNUSUAL FORTIETH:
ARVIAT AND BAKER LAKE

People in the South—and whites in the Arctic—who decry the passing of the old ways and customs in the North forget one vital factor. Eskimos are human beings. They don't want... to live in a crummy snow house, to be bitterly cold and half frozen and hungry most of the time. If they can make a living in a settlement, by casual employment, by carving, perhaps by running a few traps on the side, then most Eskimos will accept this chance and even seek it.

DUNCAN PRYDE[1]

SEPTEMBER 2003 Fortieth wedding anniversaries are special: they speak of enduring love, but also of a lot of hard work. On our part, we jointly decided to recognize this signal achievement with a trip to Arviat and Baker Lake (Qamani'tuaq) in 2003. Many discussions resulted in an agreement to use our spare time and resources to focus on the Inuit and their lands. Going elsewhere would be a distraction and an opportunity lost. The joy we experienced meeting Inuit in Pang remained fresh in our minds. Many of the surviving artists represented in our collection lived in Arviat and Baker Lake.

By September, people would be back in their communities after a summer out on the land, and the flies and mosquitoes would have abated.

Our friends were mystified about where we were going, and why. One friend dubbed this a "vacation to hell," reflecting his own tastes, of course. Whenever we mentioned we were going to the Canadian Arctic, friends almost always responded "Alaska," as though this were the only place in the Arctic. This is truly terra incognita to most people, which further added to its appeal.

What our friends did not understand, but we were coming to realize, was that we were on a quest, seeking the simple and more natural life that the Inuit represented. Enmeshed for most of the year in an often stressful and highly commercialized world, we yearned for something more idyllic. Although trips such as this were packaged in convenient bits of time, we were thinking about them all the time, reading, planning, and were in the process of a change in how we saw ourselves and the world around us. In a very real sense, the Inuit represented a focal point in our combined journeys of self-discovery and self-renewal. Something deep told us that we needed to get to know the Inuit, to experience their land, to immerse ourselves in their culture. No guides of any sort for this trip; we would trust in our ability to meet people after we arrived.

The only commercial flights to Baker Lake and Arviat flew from Winnipeg, enabling us to visit Faye Settler's Upstairs Gallery. Faye was one of the premier Inuit art dealers and relished talking about her earlier visits to Baker Lake. She told us of an exhibition of Marion Tuu'luq wall hangings that was on display at the Winnipeg Art Gallery (WAG) nearby. Tuu'luq lived as a hunter-gatherer, traveling nomadically from igloo to tent her whole life, only coming to "town" in Baker Lake in the 1960s when she was about fifty years old. Her work was informed by her skill in sewing skin clothes, but obviously drew on a deep need and uncanny ability to combine color and form in original ways, using wool cloth and embroidery floss as her medium. Frequently abstract, always full of glorious color, and always original, her art was humorous at times, but also offered profound commentaries on life.

Some pieces were huge, and others small; some were made on the land in a tent, and others in a tiny prefabricated home in town. When curator Marie Bouchard showed Tuu'luq photos of these wall hangings in her home in Baker Lake, Tuu'luq cried, and then complained that Bouchard should have come sooner when Tuu'luq could still see![2] Before the exhibition opened, Tuu'luq died. Experiencing her art was a perfect starting point for our visit to Arviat and Baker Lake.

A Calm Air twin-engine turboprop carried us over the marsh and scrub trees of northern Manitoba to Churchill, which we visited for a couple of days. Many authors describe modern Churchill in derogatory terms, because of the scrap heaps from decades of army use, particularly during the years of the Cold War, and because of the crude, rough people who gather in Churchill. Nonetheless, this is a place with great natural history. White beluga whales calve by the thousands in the Churchill River, usually peaking in July and August. Churchill is most famous for the polar bears that come through by the hundreds every year. The bears follow an ancient path from their dens on the shores of Hudson Bay to a peninsula nearby where the ice first forms each winter. Tourists flock here especially in the summer to see the whales and in October and November to see the bears, or in deep winter to see the Aurora Borealis. We chose our time for this trip to avoid the tourists.

No one was at the little air terminal in Churchill to meet us, and no one answered the phone at the Polar Inn when we twice called, even though we had a confirmed reservation (with encumbered MasterCard). While waiting, we marveled at an immense polar bear towering over us behind glass in the airport building, a frightening aspect even though he was forever immobile. Soon we were the only people remaining in the airport. Luckily we found someone who would drive us into town. The driver dropped us at the Polar Inn, but it was locked. Later we learned that the owners left for a little vacation after the beluga people departed and before the bear folks came. Traveling light is a good thing; we carried our bags across the dirt road and got a room in the Churchill Motel. It was cheap, and we knew from experience that one usually gets what one pays for. We dragged our stuff up a slanting plywood floor covered with a dirty rug through a narrow, dark, stale-cigarette-smoke-filled corridor to a small room. The rear view looked onto what appeared to be a metal scrap yard. The restaurant was closed. No one else was there besides the proprietors. We took a walk around town, and found a place to eat something. After we went to bed, we heard a siren signifying curfew (too many thefts by young people) and a gunshot close by darting a bear that was in the same area where we had just been walking. The next day we were cautioned about wandering the streets; we were told about a guy whose arm was torn off by a bear sometime in the past, and another who literally lost his head one night to a visiting polar bear.

The town is located next to the Churchill River, where it empties into Hudson Bay. It is small, with only thirteen hundred people, made up primarily of white people, plus some Indians and a few Inuit. There is a classic old railroad station and a massive grain elevator, both the result of ambitious but foolhardy plans to ship Canadian grain by rail out through Hudson Bay. The problem was the waters were free of ice only from July through October. Scrub spruce trees were common in and about town, since the tree line is a little further north. There were lots of empty tundra buggies, hotels, and motels, awaiting the influx of bear watchers with their cameras. Snow geese were everywhere, even marching down the railroad tracks. One restaurant was open, and we talked to some hunters and local people who had been up north for various reasons for variable amounts of time. They were amazed that we were headed to Arviat, which seemed of no interest to anyone. "Why would anyone go to Arviat?"

A visit to the small Oblate father's Inuit museum was memorable for the excellent Inuit ivory carvings and artifacts collected by the Catholic missionaries over the last century. Because long walks seemed unwise, we rented an ancient Ford truck, with 160,000 kilometers (100,000 miles) on it. Bouncing along the isolated roads, we saw stunted forests, clear lakes, ducks, and occasional trash piles. The Wild Dog Bar was tucked into a secluded corner not far from an old, abandoned, former Cold War radar station, now gone to ruins. A very rough and mean-appearing woman stepped out from the bar. Reading her animosity, we were not tempted to go in for a beer. A ship in ruins was not far from shore, and the wreck of an old cargo plane was just outside town. Small, lonely spruce grew from fissures in great slabs of rock by the shore's edge. Hudson Bay looked immense, and very cold. We drove to the enormous town trash dump and landfill in the evening, hoping to see bears prowling around, but were disappointed.

Not everything was depressing. The shores of Hudson Bay were pastel pink and gold in the evening light. Tundra plants in their deep red fall colors clustered around the gray granite rocks of the Canadian Shield. The mouth of the Churchill River and the Prince of Wales Fort were close. The fort dates from the early eighteenth century, a reminder of the French and English conflicts over rights to the rich fur trading territories to the west. We were alone and in bear territory while roaming the cliff above the river. The skies turned rose and then a very deep, intense orange-red, casting a red glow onto the water. Hundreds of white beluga whales were in the bay and returning to the river, swimming slowly upstream, softly whistling, hissing, and blowing. The waters turned fiery red, embracing us with color, rendering us silent and motionless. Big white tundra swans were at a distance on the water, and hundreds of white snow geese with black-tipped wings were close by on the hills and overhead as they took off into the sunset. We wandered and took photos, trying to capture essence, spirit, purity. Arctic hares and Arctic ground squirrels ambled amid the rocks. A beautiful little cemetery was framed in the gold and red of the sunset as we returned to town.

If the refuse and the dumps and the derelict remnants of the Cold War were signs that our friends were correct in dubbing this a vacation from hell, there was heaven here as well. These glorious skies have special meaning to the Inuit. Igjugarjuk, a shaman from the Kazan River area northwest of here, told Rasmussen, "A youth is dead and gone up into the sky.... And the Great Spirit colors earth and sky with a joyful red to receive his soul."[3]

Arviat was only a short Calm Air hop up the coast. On the way, we soon left the province of Manitoba and entered the territory of Nunavut, the land of the Inuit. Trees were replaced by lakes, rivers, and swamps, extending to a limitless horizon, the fabled Barren Grounds. Nunavut consists of about 750,000 square miles of almost entirely treeless wilderness, almost three times the size of Texas, one-fifth of the mass of Canada. The Barren Grounds occupy roughly one-half of this territory, the rest being composed of Baffin Island and many of the islands of the Canadian Arctic archipelago, including Devon and Ellesmere islands. Measured another way, it is estimated that there are 300 million acres on the Barren Grounds. The Barrens are a sort of wild kingdom, and for some represent a state of mind as much as a specific place. Although as much as 40 percent of the land is covered in lakes and rivers in the summer, the area is relatively arid, with only ten to twenty inches of precipitation annually. The summers are short and cool, and rates of evaporation are slow; the permafrost (permanently frozen land) and hard bedrock of the exposed Canadian Shield prevent water from seeping below the surface. Only slight hills perturbed the flatness of the landscape; once long ago this area was mountainous. Time and the grinding of the glaciers leveled everything. Some of the oldest rocks on planet Earth were discovered near here; the world's oldest rock was discovered on the eastern side of Hudson Bay.[4]

Arviat—the name comes from the Inuktitut for bowhead whale—once was a camp for the coastal Paallirmiut. The community consists mainly of prefabricated houses for the largely Inuit population, now two thousand strong and growing. The town seemed very poor, with lots of stuff lying around. They say it looks a lot better in winter when everything is covered with snow. We stayed at the Inns North, which had an unfortunate policy of charging a married couple a two-room rate. Apparently we were the first tourists in at least ten months, excluding hunters. The manager of the lodge seemed as surprised that we had come to Arviat as people in Churchill were surprised we were going to Arviat. Our companions were a Canadian dentist who was in town for a few weeks, and an official with the housing authority who was trying to get the local communities to adopt more energy-efficient designs in their construction. He was as sympathetic with the Inuit as the dentist was unsympathetic—perhaps because the dentist was appalled by the sweets/cigarettes/lack of tooth brushing/rapid decay that he found so often. There also were some construction guys and some big, very fat, loud American hunters.

Two years ago, on our first brief visit to Arviat, Joyce enjoyed the company of Mary, Lucy Tasseor Tutsuitok's daughter. Joyce wrote letters to Mary, sent her pictures, and alerted her to our arrival. We let it be known on arrival at the inn that we hoped to meet Mary. The word spread fast, and we were soon introduced to Mary Tutsuitok, a lovely young woman, but not the same Mary whom we met previously. We showed pictures of Mary, which cleared things up. Mary Tutsuitok was married to a brother of the Mary we knew. "Our" Mary was married to Peter K., the mayor, and went by her married name, thus the confusion. Mary K. was extraordinarily kind to us, understanding our interests in the town and its well-known artists, including her mother.

Mary and Peter and their six kids lived in a typical, small, one-story frame house. There was a large anteroom to take off shoes, store winter clothes, and buffer the house from the winter cold on entry and exit. A living room and a kitchen were the main gathering places. A piano keyboard was in almost constant use by the children. Housing and utility costs were heavily subsidized and very cheap by our standards, but everything else was expensive, including gas for skidoos and ATVs, food in the Northern store, cigarettes, candy, and soft drinks for the kids. Peter won the mayoral election last year, and was now out of town attending a mayors' conference in Rankin Inlet. Before the election, they depended in significant measure on his hunting ability and on Mary's skill in cleaning and preparing the caribou, beluga, and seals he shot. They were happy for the steady cash income his mayor's job provided.

The community happened to be at the onset of an annual two-day celebration, "Hamlet Day." Festivities were under way in the gymnasium. Country food (beluga, char, caribou) was lying on papers scattered on the floor, and Inuit of all ages ate the food with abandon. An elder approached and offered some caribou. He opined that the young were killing too many animals because there were not enough elders to provide guidance. There were group games, first a sort of musical chairs in which we participated after the master of ceremonies called out to us over the loudspeaker, "Hey, you white people, get into the game." We were enthusiastically received, and had a lot of laughs. The acting mayor told a long story about a man from the community who was being honored for his efforts that saved a man from death after an accident on the ice. We were met with big smiles by the great stone sculptor Luke Anowtalik, a tiny, wizened old man, as well as Jacob Irkok, another well-known carver who specialized in caribou antler. If only we spoke Inuktitut!

Shortly, the acting mayor approached us and invited us to be judges to select Miss Arviat. What a frightening prospect. How would we judge? Would we embarrass them and ourselves? What criteria were appropriate here? Our fellow judge was a local woman who fortunately spoke Inuktitut — fortunately, because some of the contestants prepared a speech that was given in Inuktitut. Five young women were to be judged on the appearance of their costumes and the message they delivered, about raising community standards and self-esteem. Some of the contestants were terribly shy and scared, and they spoke so softly we could not hear them over the din of the little kids running around. One of the candidates was in college in Ottawa and delivered her talk with great assurance. She was attractive, attired in beautiful homemade sealskin kamiks and caribou skin amautik. Our fellow judge suggested that she should be the winner, much to our relief.

The Northern Lights were on display as we returned home. Soft translucent blue greens spread over the top of the night sky, flickering and shifting back and forth, creating a slow dance on a majestic scale, a gift from a far distant solar flare, ionizing the plasma high above earth. One of many traditional Inuit beliefs about these colorful streams of light across the skies is that they represent a game of football played by spirits of the dead who live in the heavens. The Northern Lights were a force to be feared, as related by legends of Inuit who lost their heads to the long reach of the Aurora. Those heads became balls for the games of football. Whistling might make the strange lights disappear, or one might hide from them inside.[5] We whistled, and stepped inside. The time of twenty-four-hour sunlight was past, and the long, dark days were fast approaching.

Mary took us to see her mother Lucy, whose house was nearby. We wanted to see Lucy for several reasons, including a chance to hear her interpretation of her 1990 carving of a sad mother walking away from a small child. This is an unusual Lucy piece, and was carved at about the time Farley Mowat's book[6] circulated through Arviat, telling the tragic story of Kikik's trial for murder of her abandoned child. Although Kikik was found innocent, the story caused much anguish in the community, after being suppressed for decades. Kikik and her surviving children lived in Arviat and certainly were well known to Lucy.

Lucy was seated on the ground in front of her house, using a power tool to carve three gray stone pieces. The stone here is particularly hard, and difficult to work.[7] Her earlier work was done with a hatchet and files. Lucy had a deeply lined and weathered brown face, prominent cheekbones, a flat nose, and bright dark eyes. She had just adopted a small child in diapers, and was thrilled with her new duties. She seemed quite energetic. We showed her a picture of the carving in question, and she recognized it immediately and with delight. She informed us through Mary that the carving showed a female shaman leaving behind her frozen husband, whom she had tried unsuccessfully to revive. Similar to what we thought, but not Kikik. Or so she said. More experienced experts have told us the artists frequently change stories regarding their art and its meaning. Moreover, Inuit may respond to direct questions with purposely inaccurate answers, as though to teach the rude person who asked the question some manners. It is better in Inuit culture to learn by observing than by intrusive personal questions. We later learned that the people of this hamlet tried to keep Kikik's story secret from her surviving children, so we have grounds for believing her sculpture recalls the Kikik story. At the least, when we look at it we are mindful of the terrible privations that Lucy and her people suffered, not so long ago.

64 CHAPTER FIVE

FIGURE 11 Lucy Tasseor Tutsuitok carving a statue outside her home in Arviat, 2003.

FIGURE 12 Lucy Tasseor Tutsuitok, *Mother and Children*, c. 1975, Arviat. An early Lucy piece carved with hand tools, showing the importance of mother.

Mary was extremely gracious. She served us tea, and gave us a luncheon meal with soup and a caribou-and-rice stew. Hearing that it was Fred's birthday, she quickly drew pictures of caribous, seals, and bears, a much appreciated present. She arranged for her oldest son and her tall and striking stepson to take us out on the tundra with their Honda ATV, which we did with delight on a warm day. After a few miles we had a flat tire, but fortunately some other folks were going hunting on the same long gravel road, and they fixed it. We drank on our knees from crystal-clear streams. After bumping along an endless number of tundra miles with both of us perched on the back, we arrived at a low hillside covered with little black crowberries. Twenty-five years ago, somewhere on the tundra close to here, Elizabeth Nutaraluk carved a remarkable abstract female shaman with caribou antler eyes, teeth, and hands imploring the heavens for help, now an important part of our collection. A tundra hawk soared low over the vast open tundra, and sandhill cranes competed with us for berries, staying just ahead of us. The boys were much faster than we at finding and picking the berries, and filled their plastic bags much more quickly. We returned with berry-stained pants and quarts of black, small berries, which all of Mary's kids devoured as a special treat.

That night we returned to the community center for drum dancing and singing. The participants were primarily elders, and the women were attired in sealskin kamiks and decorated white cloth amautiit. Some of the men were in traditional attire, others in modern clothes. The women huddled together on folding chairs, while one man after another danced in the center. The rhythm was slow, and to our untutored ears the songs sounded similar. There was a different song for each male dancer; traditionally, each man owned his song, which was passed along by oral history. The women joined in on the refrains. They were doing this entirely for themselves, and we were just fortunate observers of something that was very common in traditional camp life, often in special large igloos made for feasts and celebrations, or for resolving conflicts through songs of derision. In the old days, song fests could last for many hours, and sometimes lifted the group into states of ecstasy.[8]

Rasmussen observed many such celebrations in his travels among the Canadian and Greenlandic Inuit, and recalled the power of the performance later:

> And now, when I remember the inexplicable way in which the words, music and dance mingled into one great wave of feeling and for a moment made us forget everything else, I can understand more clearly than ever, how difficult it is to take the songs of the Eskimos out of their own context. For the words of the songs are only part of the whole intended effect. Read an opera libretto without music, staging and performers, and you have a comparison.[9]

We would have loved to be able to translate the songs, but could imagine their content based on the many songs and poems we read in the writings of James Houston and Knud Rasmussen. Perhaps one of them sang something similar to one of our favorites, easily appreciated even in the absence of drum dancing; poetry in any language.

> *The great sea has set me in motion*
> *Set me adrift*
> *It moves me as the weed in a great river.*
> *The arch of sky*
> *And mightiness of storms*
> *Encompasses me,*
> *And I am left*
> *Trembling with joy.*
> INUIT SONG.[10]

Mary insisted on dressing Fred in Peter's caribou parka and sealskin boots. The coat was much too large; Peter like many Inuit men is thick and strong. At least the sight of Inuk Fred was very funny for the ladies. We returned to

Lucy's home and picked up a finished carving we had selected the previous day for our son, Brad, and his attractive wife, Suzie. It was light gray the first day but a dark gray the next day, after being oiled and cooked in a frying pan. Later Mary told how her mother started carving years ago: they were desperate, out of money and food, and Mary heard her mother crying. Lucy had to start carving to earn money. She produced some fine abstract work, considered by experts the equal to John Pangnark, a contemporary of Lucy's in the early days of Arviat carving. Her work often depicted simple heads of people emerging from rough hard rock, symbolizing the solidarity of families and their connection to the land. Pangnark developed a highly abstracted style that emphasized planes and geometric forms, and was compared to the great sculptor Brancusi.[11] They knew they were good and delighted in informal competition. Her best work now brings large sums at auction, but unfortunately she receives little of the financial rewards. Pangnark is long gone.

Martina Anoee, one of the singers at the drum dancing the previous night, received us in her home with great enthusiasm. With the help of Mary's translation, we had a marvelous visit. Martina makes world-class seal- and caribou-skin dolls (*qamukaq*), with caribou leather faces that resemble the faces on American apple-face dolls.[12] We commissioned a female doll, to be picked up at the airport on our return. Suddenly, to our amazement, Martina started to teach Joyce to throat sing. This ancient skill is hardly singing to our ears, more like a guttural chant straight out of the esophagus or stomach. Singing is done at very close quarters, faces inches apart, the goal being to see which woman starts laughing first.[13] Joyce secretly hoped she would have the opportunity to learn throat singing and tried gallantly, which Martina appreciated. Both laughed lustily, along with a visiting friend from Baker Lake. The throat-singing instruction went on for quite a happy while, a special moment. No one was less important or more ignored than Fred, a lone male in a group of strong women.

Two days passed quickly. Before we flew to Baker Lake, Joyce gave her silver earrings to Mary's oldest daughter, who had pierced ears and wore the earrings proudly. She returned a gift to Joyce, the first pair of beaded mittens she had made.

Baker Lake or Qamani'tuaq is "the place where the river widens," a very wide place in the Thelon River, nearly fifty miles long by ten to twelve miles wide. The hamlet of Baker Lake is close to the geographic center of Canada, and is in the heart of the great Canadian Barren Grounds, two hundred miles west of Hudson Bay and just a bit below the Arctic Circle. It is the coldest place in mainland Canada, with temperatures every winter of at least $-50°$ F ($-46°$ C), often with gale-force winds. There is no buffering by Arctic seas, and the cold from the high Arctic sweeps down unimpeded. Temperatures as cold as $-80°$ F ($-62°$ C) have been recorded here. The closest tree that looks like a tree is about two hundred miles south. The lake is connected to Hudson Bay through the long reach of Chesterfield Inlet. The hamlet sits at the head of the lake, near its confluence with the Thelon and across from the other great river that flows into this lake, the Kazan. The Thelon comes in from the west, its strength supplemented by inflows from Dubawnt Lake and the Dubawnt River. The mighty Kazan comes in from the south. The Thelon and the Kazan, along with the more northerly Back River that flows east into Chantrey Inlet and the Arctic Ocean, help delimit much of the territory of the inland Caribou Inuit. We were to spend many weeks in the years ahead paddling these rivers.

The little airport building displayed a number of Baker Lake prints, an artistic welcome for us who knew no one in the community. We were met at the airport by Elizabeth Kotelewetz, who, along with her husband Boris, ran the Baker Lake Lodge. In the soft evening light, we could see homes and buildings dotting the shore's edge. The old Anglican church, where missionaries established a base almost eighty years ago, was prominent by the side of the lake. Rasmussen passed through here on his Fifth Thule Expedition. Our room at the lodge was the former jail in the old RCMP building. For a jail, the accommodations were mighty nice, although our experience in jailhouses was minimal. We shared an indoors bathroom with another couple who

FIGURE 13 John Pangnark, *Figure*, c. 1970, Arviat. Pangnark was a master of the abstract figure. Note the barely discernible face.

FIGURE 14 Mark Alikaswa, *Men Fighting Over a Seal*, c. 1958, Arviat. Alikaswa was also a great carver.

occupied an adjoining "cell." This was the season when caribou were fat and had the best skins, and a hunting party from the United States was in town. Although we could not identify with them, the hunters provided financial remuneration for the Inuit, whereas we brought only lots of questions.

The main building of the lodge contained scores of photos of Inuit from the community, some of whom we recognized from books we had read. There was Luke Anguhadluq, unmistakable with his tall, lean frame, and great shock of all-white hair. He was a well-known print artist who had supposedly not used paper and pencil before he and his family made the final long walk of about ninety miles over the tundra from the Back River area into Baker Lake in the 1960s.[14] Anguhadluq foresaw the effects of entering a white culture, and tried to prevent his family from learning English. His second wife was Marion Tuu'luq, the wall hanging artist. Hanging on the wall was a picture of Anguhadluq's cousin and very close friend, Jessie Oonark, probably the most famous artist from Baker Lake.[15] Many of Oonark's children also were famous artists, as were some of Anguhadluq's. A photograph on the wall showed Tuu'luq, Oonark, and Anguhadluq in front of a tent, working on Tuu'luq's *Lake Trout*, an early and great wall hanging.[2]

Anguhadluq, Tuu'luq, and Oonark came from the small band of Utkuhiksalingmiut, who lived north of here near Chantrey Inlet and the Back River. Rasmussen visited them in 1922 in his travels; he called them the cleanest, most handsome, and least contacted by whites among all the Inuit whom he visited in his three years in the Canadian and Alaskan North.[16] Most of them had not previously seen whites. The Utkuhiksalingmiut were among the last to come into Baker Lake. Many were forced off the land by the failure of the caribou migration and attendant starvation in 1958. Others remained in their tundra camps for another decade. The anthropologist Jean Briggs lived among them from 1963 to 1965, and recorded the customs and manners of the people; great emphasis was placed on maintaining equanimity and good cheer in the face of adversity.[17] They were social outcasts when they arrived in Baker Lake, staying in tents and igloos. Elizabeth Kotelowitz told us that when she arrived in the late 1960s as a nurse, the people from Chantrey Inlet and the Back River were treated as low caste by other Inuit, because their English was poor and they were not acculturated to the ways of the white man. Often Elizabeth observed them waiting until the end of the day to be seen by health workers, regardless of when they arrived in the clinic.

The Utkuhiksalingmiut were one of nine separate groups that eventually were relocated to Baker Lake. Some were from the south along the Kazan (including the Ahiarmiut, Paallirmiut, and Harvaqtuurmiut, all Caribou Inuit), others from near the Thelon to the west, from the north along the Back (the latter included the Utkuhiksalingmiut, members of a larger group often designated as Netsilik Inuit), or from the east closer to Hudson Bay. All shared dependence on caribou and fish for their yearly survival, although the Utkuhiksalingmiut and those from Hudson Bay also hunted seals, particularly in winter. Once they got to Baker Lake, seals no longer were an option; they depended on caribou, moose, geese, bird eggs, trout, char, and the occasional musk-ox, plus food brought in from the outside by yearly visits by a barge or by irregular air transport.

Our cabinmates were Shaun and Marcella, a gold prospector and geologist, respectively, employed by a mining company. Their helicopter pilot joined us at the table, and all three related fascinating experiences. They had seen Arctic wolves recently, and everyone said there were many musk-oxen and caribou in the area. While we were there, Shaun discovered gold, barely containing his excitement. Finding gold was not necessarily good news, since it foretold a potential new mine somewhere close. Some view gold and diamond mines as the future salvation of Nunavut; Andrew Qappik's design for the new Nunavut flag has a prominent gold section, signifying the wealth of gold within the new territory. Others view the mines as the potential ruination of the land and its animals.

Shaun was a passionate loner and lover of the wilderness, and had camped throughout this and other parts of Arctic Canada in many decades of prospecting. He rapturously described seeing the Northern Lights when he was camping in the mountains of the Yukon Territory, seeing beautiful red and green colors cascading over the snow-filled slopes and valleys. Having camped on the lakes of the region south of Baker Lake, he knew Kikik's camp on North Henik Lake. He gave us detailed maps and instructions as to where to pitch our tent when we got out on the land alone, and what to bring: "a shotgun is good for bears, but remember, two shots." He advised staying alone in one place for a couple of weeks to get a feeling for the land, and to get in tune with ourselves. A float plane could drop us off and pick us up. Mosquitoes and black flies might be bad, and bears might be a problem, but it would be a good experience. It was tempting.

The town was more orderly and appeared more prosperous compared to Arviat. Almost all of the thirteen hundred people are Inuit. The young and middle-aged all speak English, and the young are said to be losing the ability to understand the nuances of Inuktitut. The former center of printmaking, the Sanavik Co-op, now was a grocery store. The Jessie Oonark Center no longer made prints, but produced souvenir T-shirts and ties, and sold locally made silver and ivory jewelry. There were several churches; the usual Northern store with an attached fast-food joint (Kentucky Fried Chicken and Pizza Hut); another Co-op; a couple of Inuit-run small hotels; a health center; a new, large secondary school; an Arctic College outpost; and a home for elders. At the Heritage Center we found a book by Hattie Mannik, containing transcribed and translated stories by elders who grew up on the land,[18] saving their lore for future generations.

David Ford, a good friend of Faye Settler, ran a fine little art gallery. We bought two small carvings, one by Miriam Qiyuk and the other by Josiah Nuilaalik, each of whom was a child of Oonark and grew up on the land in the traditional way. David insisted we take his automatic-transmission ATV for travels around town. It was only a short ride to Luke Anguhadluq's grave, which sits alone in a place that once must have been beautiful, but now rests very close to the town sewage treatment pool. A lone Arctic hare sat still, hiding among the rocks in his not-yet quite white coat.

Our initial reception in town seemed cool relative to Arviat, where everyone smiled on first glance. Once we asked questions or introduced ourselves, however, the response was always warm and helpful. We started at the Heritage Center, and got hints as to where to meet people who could introduce us to the artists whom we wanted to meet. So instructed, we went to the Co-op, looking for Hanna, daughter of Ruth Qaulluaryuk Nuilaalik, a great wall hanging artist and daughter of Luke Anguhadluq. Ruth walked in off the land forty years ago with her father. Hanna was at the computer, checking people out. Joyce approached her and asked if she were Hanna. She literally jumped backward, and looked in amazement at us. Later she shared that she was afraid when we introduced ourselves, thinking that if white people knew her name, she must be in some sort of trouble. She beamed a beautiful smile and agreed to help us.

When Hanna took us to meet her mother, we discovered that her husband was Josiah Nuilaalik. Thus at a single visit we met a son of Oonark and a daughter of Anguhadluq, and a stepdaughter of Marion Tuu'luq. Ruth and Josiah, seventy-one and seventy-five years of age, respectively, were alert, short, slim, and healthy appearing. We visited them several times. Inside their tidy home, we met members of their family and conversed through Hannah with Ruth about her works on cloth, or wall hangings. She had no pictures of her work, but we discussed three of her landscapes pictured in *Works on Cloth*, a Marion Scott Gallery catalog, that we had brought along with us. Joyce was very impressed with Ruth's quartet of wall hangings titled *Four Seasons*, now held by the Winnipeg Art Gallery, as well as others displayed in various exhibitions.[19] Surprising Fred, Joyce proposed that Ruth do a wall hanging for us, on commission, to which Ruth readily agreed if we could get her good-quality embroidery floss and an old-time thimble. Previously, Joyce declared that we were not going to collect another wall hanging, because they are too large and take too much wall space. Evidently Ruth was inspirational, and the deal was done.

While this was transpiring, Josiah came in and saw his little statue that we purchased at David Ford's Ookpiktuyuk Art gallery. Hanna told us he was not pleased with the caribou antler on that piece. Josiah went outside again, we thought because he was not interested in Qallunaat talk. After an hour, however, he reappeared with beautiful small carved antlers for our little flying shaman, and showed us with a flourish and a huge smile how the shaman flew through the air.

Josiah talked in Inuktitut about many things, including a good-natured but somewhat off-color story that involved Ruth when she was in Baltimore for *Northern Lights*, an Inuit wall hanging show. Poor Hanna clearly was embarrassed doing the translation. Ruth listened very intently, with a fierce expression. Obviously she understood the story in Inuktitut, but she waited for the translation to finish before she broke out in a huge smile and a big laugh. Watching her, we laughed also. We were amazed at the way they accepted us into their lives so quickly. Who were we, strangers from somewhere south? But there was pleasure in sharing the moment. Tea and bannock added to the pleasure.

Hanna's brother told us with a straight face that the char in the Back River were so huge that he could carry only one on his back at a time. Perhaps our Toolooktook carving of a hunter with a fish, depicting an overwhelmed man carrying a huge fish, is not apocryphal. Indeed stories of huge Back River char are abundant in several of the tales of Utkuhiksalingmiut elders.[18] Briggs observed the fabulous fall migrations of salmon trout that typically weighed ten to forty pounds.[17] The large musk-ox horn and wooden *kakivik* (fish spear) lying out back

FIGURE 15   Josiah Nuilaalik, *Fox-Bird Shaman*, 1991, Baker Lake. Elegance in stone.

of the Nuilaalik home were further evidence of the size of the fish. When Hanna described how as a child she was afraid of the giant char in the Back River, we eventually believed her. The native name for the river, of course, means "Great Fish River."

Ruth is an ardent fisherwoman, going out on the ice in Baker Lake at any temperature to seek and sometimes find big fish of the sort they described up north on the Back. When we inquired of others about her, they often referred to her as the old woman who is always out fishing in the cold. Hanna later told us how thrilled Ruth was when she caught a giant lake trout so big that she could not get it through the hole she made in the ice, and only landed it with help from others who saw her plight.

At the Oonark Center we met the manager, Paul, the grandson of Jessie Oonark and son of Miriam Qiyuk. Paul took us to meet his mother at her home. Miriam was there with her husband, Silas, who made sure we knew he was as well known as his wife. Silas carves, and Miriam is one of the best of the contemporary wall hanging artists as well as a fine carver.[20] Miriam had some raw *muktuk* (*maktaak*)—beluga whale skin and fat—beside her on a plate, and brought it to our attention without offering it directly. It undoubtedly was bad manners on our part not to ask to taste it. Knowing how to behave was a bit of a mystery. It was hard not to be or to feel rude.

Irene Avaalaaqiaq made the one wall hanging in our collection at that time. Her son Basil told us she was out hunting. Tomorrow would be better. We met in her home, where we found her coughing and fatigued, but dramatic in appearance, a large woman with smooth brown skin and white hair. She came to life when we showed her a picture of her wall hanging that we purchased in Rankin Inlet in 2001.

She explained it to us, Basil translating. In the dark of night (blue background), an abstracted white wolf is in the process of transforming into a human, and is exhorting fantastic yellow ptarmigans around him to do the same, some of which are doing so. The red forms on the border are rocks in the process of transforming into birds or humans. We wondered whether the

FIGURE 16 Irene Avaalaaqiaq, *Untitled*, 2001, Baker Lake. Wall hanging of the transformation of animals and rocks.

imagery was Christian, but most of her work was based on stories and myths learned from her grandmother while they were living in skin tents and igloos on the tundra around the lower Kazan River.[21] She grew up with no other children, and was told to imagine friends with whom to play.

Walking was a means to meet people on the road, and we often spoke with smiling Inuit. One middle-aged man seemed entirely lost as he frankly admitted to having been an alcoholic, and having wasted his high school education. Having been south, he knew the cities were not a good place for him, so he returned to Baker Lake. He was half white and half Inuit, had no extended family, lamented not knowing how to hunt, and had little hope of a job. He complained of being extremely bored, and mentioned that the future does not seem any brighter for him. Another man drove his old car back and forth on the main road, seemingly just to pass time. Coming upon a caribou feast at the edge of town, we were welcomed immediately and asked to participate. Bannock and caribou ribs from a boiling stew were delicious, but we foolishly passed on raw caribou, which was being carved by old ladies with ulus, the woman's knife. We were carefree, and loved soaking up the tastes and smells and pace of the hamlet.

Jessie Oonark's grave is on a ridge about four miles out of town. It seemed a long trek, but we had time to wander. We walked at first along the gravel road to the airport and then across the tundra. The colors of the tundra were fabulous, and one could see immediately where Ruth Qaulluaryuk drew inspiration for her wall hangings. The turf was spongy and moist in places, dry and rocky in others. Joyce collected caribou antlers that she cherishes to this day. Sometimes the tundra was remarkably fragrant, from stepping on Labrador tea plants. There were many types of little berry plants, including crowberries, bearberries, blueberries, and lingonberries (mountain cranberries), all turning varying shades of red.[22] The lichens were of many types and colors; some were white, very pretty amid the reds and purples and greens and yellows of the heath plants around them. We climbed a long incline up a ridge covered in shades of red, pink, and purple. Oonark's grave on the top of the ridge was a simple pine box covered by rocks, on top of the permafrost. The land here is permanently frozen to at least seven hundred feet, based on findings in Rankin Inlet when tin mines were dug into the earth. A cross with Oonark's name on it marked the grave as hers, but many of the letters had fallen off. Oonark's spirit looked a long way across the tundra in every direction, and could see Luke Anguhadluq's grave on the other side of town. They planned it this way. Each of them chose to be buried alone on the tundra, away from a community graveyard. The Thelon flowed hard below us, and far across the lake, the Kazan poured its waters into the lake. No caribou were in sight, but they had to be out there somewhere, since they always return to feed near town about this time. A winding one-lane dirt track took us back to town, an easier but less satisfying walk than our path across the open tundra.

The flies and mosquitoes were entirely absent, even though there was a heat wave while we were there; most days were about 50°F (10°C). Usual temperatures were close to freezing. Soon winter would come. It was close to time for us to head for home, and we grew contemplative. What will happen to Hanna and Basil and the others? Will good-quality art continue after the remaining elders pass? A walk along the shore in the pastel light of dawn before anyone else was up was soothing, and then it was time to fly home. But the journey was not quite over. The Calm Air flight landed for ten minutes in Arviat, where we planned to meet Martina and pick up and pay for the doll. As we landed, an ATV raced along the little road, raising a cloud of dust, undoubtedly Martina on her way to see us. Joyce talked her way off the plane for only a few minutes, but Martina did not reach the plane in time. We experienced a mixture of guilt and anxiety as we took off, leaving her behind with her doll, but should not have worried. She sent it to us through the mails, and we paid for it by return mail. Her doll was of a woman clothed in caribou and sealskins, with a baby in amautik, and dried apple-like faces made of chewed caribou skin. Mary later wrote from Arviat and told us her oldest son had gotten his first three seals on the ice, which made her feel proud.

Joyce found a collection of embroidery floss at the Hudson's Bay store in Winnipeg and also back home in North Carolina. After considerable search, she also found in antique stores old thimbles of the sort Ruth requested. We mailed floss and thimbles to Ruth. Months passed, and finally Hanna called saying her mom finished the wall hanging. "She must have been happy, because it is pretty good." In a few more days a fragile and dilapidated, inside-out Froot Loops cereal box arrived, barely held together by white surgical tape. Inside was a colorful rendition of the tundra in Ruth's typical abstract style. It consisted of a series of abutting and intersecting partial circles, each made of multiple colorful parallel cross-hatched bands created entirely by embroidery floss, with a pair of embroidered orange small appliqué birds at the top, facing each other. We put it above our bed, where it reminds us of the pleasure we had laughing with Ruth and Josiah and Hanna, and of our long walk to her mother-in-law's grave on the multicolored tundra above the Thelon. The wall hanging speaks to us every day.

FIGURE 17 Ruth Qaulluaryuk, *Untitled*, 2003, Baker Lake. Wall hanging commissioned by us in her home. The small birds may be her representation of us?

6

A SECRET RIVER: TO THE THELON RIVER

*The wolf is art of the highest form and you cannot be in
its presence without this lifting your spirits.*

MARK ROWLANDS[1]

JUNE–JULY 2004 June 26, 2004. Flying from Fort Smith, NWT, in a Cessna Caravan with three canoeing partners and two canoes lashed to the plane's pontoons, we could see dark-green, stunted spruce trees below us, an unbroken panorama of boreal wilderness. We were on our way to a tundra river that flowed into the Thelon. Walking the tundra in its fall colors outside Baker Lake last September, and standing high above the Thelon at Jessie Oonark's grave had been intoxicating. Inspired by that experience, we laid plans to paddle the great rivers that define much of the territory of the Barren Grounds Inuit: the Thelon, the Kazan, and the Back. The Thelon was first. No more cruise ships, we would get close to the land. This trip would take us into the heart of the Thelon Wildlife Sanctuary, and hopefully close to the great animals that help sustain the Inuit, and which play such a key part in their culture and art.

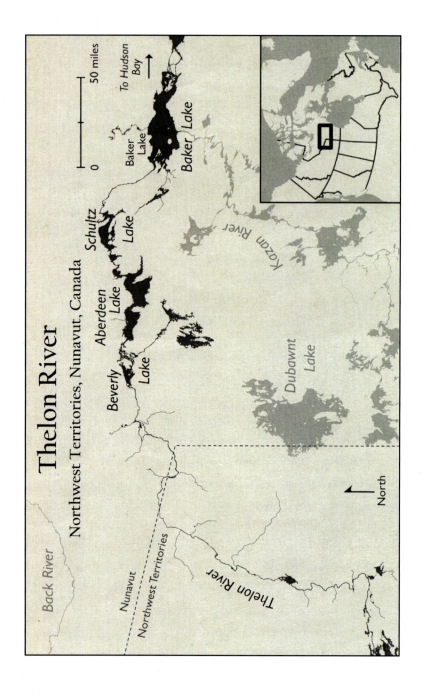

This was not a trivial undertaking. Much of the intervening year was spent in preparation. We heard tales from our friend Dale Vermillion about his three great canoe trips on the Barrens about thirty years ago, and wondered whether we could do something similar with a guide. Dale told us of Alex Hall, a very experienced guide whom he ran into years ago (1976) while Alex was on a long solo paddle across the Barrens. After searching other possibilities, there seemed little doubt Alex was the best. A full-time canoe guide to the central Arctic for over thirty years, he was the acknowledged expert on the Thelon. His wonderful book, *Discovering Eden*, describes his many river experiences.[2]

Deep roots were being tapped. We always loved the North. Joyce grew up in New England, skiing, ice skating, and literally building igloos. Fred grew up in Chicago winters, and learned paddling in northern Wisconsin. He drove up the Alcan (Alaska) highway to Fairbanks at nineteen, where he worked in a gold mine, and spent a week above the Arctic Circle fighting a forest fire in the Alaskan Brooks Range. Indelible memories included seeing caribou skulls with antlers locked in a past-death embrace, and the midnight sun behind snow-capped peaks. Both of us paddled a lot in earlier years. Fred's experience included a couple of trips on the Boundary Waters area in northern Minnesota and southern Ontario, various whitewater rivers in Wisconsin and New England, and a two-month escape to southern Quebec after his first year in medical school, where he did six hundred miles of paddling and plenty of portaging among black flies. Joyce learned to appreciate the wonders of the woods and camping at a Girl Scout camp in Rhode Island, followed by lake and ocean camps in New Hampshire and Maine. At Wavus Camps, canoeing and overnight camping trips were lasting memories. She dreamed of an Alagash River trip with the girls, but in those days only the boys were given that opportunity. Now she would make her own opportunity for an even greater trip. She loved sports and fishing for horned pout and rainbow trout in Rhode Island streams with her dad. Many years later, Joyce purchased a canoe for our new home deep in rural Chatham County, North Carolina. Surrounded by forests and mountain laurel, we paddled the Rocky River below our house, but that was only a teaser for a serious trip. The last canoe trip of any significance for either of us was a five-day family trip twenty-five years before on the Smith River in Montana with our friends, the Vermillions.

Guides help with cooking and arranging logistics, but not the individual work required for paddling and camping in the wilderness. Fred was forty pounds overweight, beset by a lifetime of asthma and restrictive and obstructive lung disease, and had been desk-bound and sedentary for a long time. Walking uphill was difficult on what amounted to two-thirds normal lung capacity. Fred asked his doctor whether he thought he could do such a trip; raised eyebrows was his response. Joyce was in better shape and better health. Though retired from teaching and conducting research on movement development, she maintained an active schedule, including golf, gardening, and woodworking. She liked to work on the land, building bridges over streams, and maintaining the gardens in addition to caring for a growing pod of ten grandchildren. It was up to Fred to get off his posterior. There is no romance in losing weight and getting in shape, so let us leave it that by eating less and exercising more he shed thirty-five pounds. Weight loss is all about discipline and motivation, and canoeing the Arctic was a great motivator. The food beast was subdued by a combination of fear and hope. Exercise led to better lung capacity; increased aerobic fitness fed into a virtuous cycle that enabled still more aerobic exercise. The asthma also was subdued. New medicines helped.

Getting into paddling shape is hard to do without putting in time on the water. The Westport River was perfect, a medium-sized tidal river next to our shared little cottage in Westport, Massachusetts. We paddled every day for a month and got on our bikes. Many days we biked twenty-five to thirty miles and canoed eight. The paddling was necessary to improve power and endurance in the big muscles of the back, the locomotives of a canoeist. We caught some big stripers from the canoe on flies. A little country store at the top end of the west branch of the river was a great place for coffee and newspaper. A paddle to a restaurant on the east branch of the river allowed a dramatic entrance and exit in our canoe, as well as fresh littleneck clams or mussels and a beer. Osprey whistled in the morning mist, warning us away from their nests. Wind and tide provided a challenge at times. The bike riding also had its pleasures, especially for Joyce. It was spring, and the roadside was adorned with azaleas, early rhododendrons, and a lot of colorful trees. Wildflowers were abundant, with wild geranium in its rose hues, lily of the valley, buttercups, and others. The air was sweet with many varieties of lilacs. Smells of manure, fresh-cut hay, and sea air filled our lungs. At month's end, sweet smells of Russian olive and honeysuckle replaced the lilacs, and the rhododendrons came out in their full glory.

Regularly scheduled flights in June took us to Edmonton, and a charter flight took us up the last leg to Fort Smith. On the way north, we saw the beginnings of the huge tar sands projects in northern Alberta, by which oil was to be released from the sands for an oil-hungry world. In the process, a great boreal wilderness was being eviscerated. Fort Smith is just north of the border with Alberta, south of Great Slave Lake, on the Slave River. In the nineteenth century this was a way station for the traders and explorers who opened up the North.

The first evening, we met our companions for the trip: Stu MacKinnon from Edmonton, a sixty-nine-year-old bachelor, lean and fit, a long-distance backpacker, bike rider, and veteran of twenty-six previous canoe trips with Alex Hall; David and his daughter Kitty from Connecticut; and Alex. David and Kitty also had substantial long-distance canoeing experience. Alex was tall and looked strong at sixty-two. The next day we met Alex's very good friend Kevin, who showed his huskies, and told us about how he and Alex hunt moose, their primary food during the long winters. A dark and smoky bar frequented by the Metis was a late-afternoon diversion. Though a novelty in the bar, we were treated well. White pelicans sat in large numbers in eddy pools, feasting on fish, in the huge, murky Slave River, a short walk from our Pelican Rapids Inn. A cougar had been seen recently just outside of town, and bears also. Voyageurs once paddled these scary-looking rapids, but few canoeists would brave them today.

Alex surprised us at the second dinner gathering, telling us that a friend recently flew over his "secret river," located 350 miles northeast of Fort Smith, and an equal distance southwest of Baker Lake, and it was totally closed in ice. This was the coldest winter and latest spring in thirty years. No one lived within 200 or more miles of our destination, and the only way to know whether the ice had gone out was to look at it. Alex was dubious that the river would be open, and had to be talked into an exploratory mission. He would fly out ahead of us in a little old Beaver, scouting to see if the secret river was open. If we were unable to land, his emergency plan was to divert south to other rivers, but they would require either a two-and-a-half-mile portage or a very long, brutal lining across boulders. Neither sounded attractive, but we preferred the latter if we must. Alex's secret river had no portages, and that is one reason we chose this trip. We were sixty-eight, and it had been forty-five years since we portaged a canoe. It was difficult back then, and it was dubious that we could do it now with the heavy eighty- to ninety-pound canoes that Alex used.

From Alex's charter flight, the views of the lakes and unbroken forests of the NWT were stunning, but we could not wholly escape our anxieties about where we would put in, and whether we could handle the rigors of the terrain. It was obvious when we crossed into the Arctic; the trees fell away and were replaced by olive and brown tundra, and the lakes and rivers were covered in ice. The boundary between the boreal forest of the NWT (the land of the Dene) and the Arctic terrain of Nunavut (the land of the Inuit) was surprisingly abrupt, created by cold winds from the far North that swoop down into central Canada in a path unbuffered by moderating ocean temperatures. Lakes freeze to eight feet deep here, and rivers to five feet. The lakes are huge, and some have ice in them all summer. It looked intimidating. The skyline was opaque from massive forest fires in British Columbia and the Yukon, far to the west and south. We passed over the Thelon, which was open, but the banks were covered in snow and ice. Alex talked with us intermittently by radio, but there was no decision for what seemed a very long time. After a couple of hours, we were close to a point of no return. A decision needed to be made very soon.

Finally we received a radio message that our intended river was open, much to our relief. We pledged to keep the name of the river secret to diminish the number of other paddlers on the same river. This river was shallow and only had enough water to be paddled successfully for about three weeks after the ice went out, and other paddlers would want to be on it at the same time. Below us were large geometric polygons carved into the tundra by frost heaves, and ice along the edges of rivers and covering the big lakes. We caught up and saw Alex unloading the Beaver below us at the edge of the secret river. After circling low over a wide spot, we landed smoothly with a big splash, taxied to shore's edge in a foot of water, and unloaded our gear, wading in ice-cold water. The day was cool, bright, and windy. Before we could say more than thanks, the planes left, and we were alone. Seeing them depart caused us to pause.

Everyone was important to the success of the whole, but Alex was most important. Alex had a ritual of rules for his clients, all committed to a tattered little notebook. The rules started out with a declaration that this was not about fun. Safety was the first consideration. He would sleep alone near the food boxes with a homemade bear alarm, so if a grizzly invaded he would be the first to know and the most at risk. He did not carry a gun, because it was better for us to be out in the wilderness aware that there was something else more powerful than we, something that could kill us. His book[2] described a few bear incidents, but only once had he had a close call, escaping neck deep in a cold pond. Reading about it was funny, but it made us aware that bears happen. We trusted that if we got into an incident with a bear, his general advice to stand still, make noise, wave your arms, and look big would suffice. If not, we would drop and assume the fetal position, protecting our private parts. We would be very careful in cooking to avoid making odors that attract bears; no pan-fried fish for us. He had a satellite phone if we needed to call in emergency help. Help for a bear attack might arrive in time to clean up the mess, but would be more useful for other medical problems.

FIGURE 18 Barren Ground camp, storm approaching.

The seats and thwarts were screwed back into tough-skinned Royalex sixteen-, seventeen-, and twenty-foot Old Town or Mad River canoes, and camp was made. Fred caught a three-pound lake trout, but foolishly put his finger in its mouth and discovered it had more teeth than a striped bass. Only by pithing the fish with his knife did he get his finger back. We had the fish for dinner in boiled pasta of some sort. The skies turned a soft, almost translucent blue-gray as the sun slowly receded, and it remained light enough to read throughout the night. There were no bugs of any sort, despite complete absence of a breeze. The tents were spacious, four-person tents for two, leaving room for our gear. We crawled into our bags, laid out on three-fourths-length, tiny inflatable mattresses, and slept the sleep of the innocent.

Perhaps Alex wanted to introduce us to the tundra before we got into the water. First thing in the morning, we walked about three miles to visit a wolf den where Alex saw a pup last year, but there was nothing in the den this time, other than old wolf prints and scat. Alex earned a master's degree studying wolves, and loved seeking them. Adult wolves are curious 80- to 140-pound animals with acute hearing of high-pitched sounds above the human hearing range. Living in packs of mainly five to eight wolves, they are loyal to and playful with family members. Their dens lie in sand under tree roots, and it was there we were looking for them. Old wolf scat is easily identified by shape and the hair and teeth within; Alex told us with a certain alarm not to pick them up lest we get a bad infection. He was referring to echinococcal disease. We really were not inclined to pick them up anyway. Joyce spotted a lone caribou grazing, and we climbed up a ridge to better see it. Alex waved raised arms to catch the interest of this thin caribou, saying how curious these animals were, and it responded by coming closer. The herd probably migrated north three or four weeks ago to calving grounds north of Beverly Lake, leaving this laggard behind to be food for wolves. A few early Arctic azaleas were in bud but not yet in bloom. The soft, white flowers of Labrador tea were all around, and were exquisitely fragrant when crushed by our footsteps. They were so dense in areas that there was no way to avoid them, nor the azaleas.

Pushing off from ice-encrusted banks into brisk currents, and getting under way at last in heavily loaded canoes, was thrilling. So much went into the preparation for this moment, and suddenly it was upon us. We felt free, and were focused only on the path ahead. Paddling was easy, and we passed snow-covered sandy eskers and dunes on both sides of the river. We were in a seventeen-foot canoe, packed with gear for eleven days, aiming to go 170 miles. A first encounter with fast water felt wonderful. We saw what we thought were two small herds of caribou on the hills, but Alex later convinced us these were moulting flightless geese, and that their apparent size was an illusion brought on by the difficulties in estimating distances in very clear light on the barren hills without trees. It did not seem possible that something that small could be mistaken for something that big. But it happened again. We were not alone in such errors; more experienced travelers of the Arctic have chased grizzlies only to discover the "grizzly" was a marmot, or have described in their field notes islands with parallel glaciers, only to learn to their chagrin that they were looking at a walrus.[3]

FIGURE 19 A wild bull musk-ox on the secret river.

After hiking up a steep hill to get a view of the beautiful, meandering blue-green river flowing between long, sandy, and rocky ridges with just a few scattered small spruce and patches of white snow, Alex spied a sleeping musk-ox in the distance. Good eyes! We paddled across the river and slowly sneaked up downwind to within thirty feet. It was a young bull, asleep on the tundra in midday. He finally awoke; we were nervous, but Alex was extremely confident. The bull shook his head, advanced tentatively on us a step or two, stopped, and considered us for what seemed many minutes. "A musk-ox will not charge a group of six," said Alex. We hoped he was right. Finally the musk-ox turned and slowly walked away toward the river with his long coat blowing in the breeze, only occasionally looking back at us.

Musk-oxen are relics of the Ice Age, and are one of only two North American mammals (North American bison are the other) to survive the great extinctions that occurred after humans appeared on the continent thirteen thousand to eighteen thousand years ago. Almost exterminated, the musk-oxen are making a comeback, here on the edge of the Thelon Wildlife Sanctuary and in scattered pods throughout Nunavut and the high Arctic islands. Although the *umingmak* (bearded one) looks something like a bison, it is a member of the goat family—a cud chewer and a climber. Standing at most five feet tall at the shoulder, and weighing up to nine hundred to one thousand pounds, their outer coat of long hair makes them look even larger. Fine fleece beneath the outer coat is said to be eight times warmer than sheep's wool, one of the secrets that allows these animals to survive without hibernating where winter temperatures of $-40°$ F are common. They roam the land looking for Arctic willow and birch leaves, Labrador tea, and sedge grass. The bull's horns have a broad base, and their horns are larger than the female's, useful in fighting wolves and other males in the rut season. Although they seem boring to some, occupying their time mainly with eating and sleeping, even in midday, they have been seen to frolic, sliding down scree slopes one after another, only to gallop off wildly in all directions.[4] Their prime enemies, other than humans, are wolves and grizzlies. They have adopted a communal strategy to thwart wolves, forming defensive circles with calves inside. Unfortunately, this evolutionary adaptation is of no use in confronting human hunters, but now they are protected. No hunting is permitted in the sanctuary, and only limited harvests are allowed elsewhere.

The wind came up at the end of the day with whitecaps on the river, and we were glad to pull in and make camp. We bathed in a shallow pond, rested, and ate a hearty dinner, yet were too fatigued after dinner to fish, even though there was a nice rise of small trout in front of camp. Our shoulders were quite sore, making sleeping uncomfortable. Hard winds rattled the tent all night and were fierce when we awoke. It was rainy and very cold. There was too much wind to paddle, so we stayed in camp. Joyce was warm in her ski gloves and parka, but Fred's five layers—including thermal underwear, wool shirt and sweater, fleece vest, and rain jacket—barely were sufficient.

A walk took us to a peregrine falcon nest, but it was empty this year. The tundra was very dry and almost brittle the day before, but after a rain it was soft and spongy, and very colorful in blacks, greens, white, and deep purple. The willows and dwarf birches were leafing out, and the willows were starting to bloom. In the afternoon we got back into our sleeping bags in the tents and listened to the wind roaring, a real storm. A few dry almonds from our stash satisfied a need to eat something. Almonds would not attract bears to our tent, would they? We read and napped. The gusts must have been forty miles per hour. The storm lasted for a day and a half.

On the second day of the storm, we looked again for active falcon nests, eventually finding one on the side of a sandstone cliff. A pair of peregrine falcons buzzed us overhead, disturbed by our proximity. On the way back we crossed an amazingly wide area covered in a maze of caribou tracks, all quite old. Obviously this was a huge crossing area at one time. The tundra takes a long time to repair damage from the great caribou migrations.

Calmer and warmer blue skies finally returned, and we paddled to a new site. Progress was slowed by seeing a pair of young bull moose in the thickets, which we stalked amid the willows. Alex called them, and again the animals responded. We were impressed that the animals were unafraid; this seemed a gentle kingdom where animals had not seen people. *Discovering Eden*, the name of Alex's book, made sense. The only animals that seemed afraid were the birds, including mergansers, scaup, and Canada geese.

A "normal" day, meaning one without bitter cold winds, graced us. The far end of a wide area was almost closed by soft "candle" ice, but there was a way around on the extreme side next to shore. The ice crackled and rang like tiny bells as we passed. We enjoyed fast water most of the way, but our confidence was tested when we hit one rock hard. We took water and had to bail, but so did everyone. Joyce paddled in the bow with draws and cross draws, keeping us out of trouble more than once. She remained calm and free of complaints, in contrast to Fred, who exclaimed and complained about the winds. We shared the chores of carrying packs and canoes from the river up to the campsites and in setting up camp. Alex admonished potential marriage partners to go on a Barren Grounds canoe trip before committing to each other. "If you want to find out the kind of stuff you're really made of, go on a long canoe trip on the tundra. If you're thinking of marrying, take your intended partner for life with you." We already had forty years in the bank, but were belatedly taking—and passing—the test. Maybe we could make another forty years.

Yellow sand dunes and eskers were all around and quite beautiful. Eskers snaked their way across the Barren Grounds in long, irregular ridges, the leavings of huge rivers that once flowed under or through the glaciers as they melted. They made a dramatic visual impact from eight thousand feet, and here at water's edge they were impressively big. In one spot there were three-foot-high ice shelves fronting a large esker. We stopped paddling to get a good look; the view was magnificent, with green canoes and paddlers in colorful parkas or jackets and caps in front of the blue-green ice, with yellow eskers towering above. Where else could one see such a sight?

Three bull musk-oxen peered out of the willows at us, as we exited our canoes to stalk them. We herded them into a little peninsula, where they

FIGURE 20 A lone wolf on the edge of the secret river.

were trapped. Before they spooked and thundered off, we got amazingly close, looking into their pupils with binoculars or a telephoto lens. One bull in particular looked malevolent and dull-witted, staring at us under a shaggy winter coat still in the process of being shed. Fortunately they skirted us rather than coming right for us, and left on the bushes patches of their very fine, downy inner coats (*qiviut*), the softest, warmest, and most expensive wool in the world. They were quick and agile.

Suddenly, while we were moving through moderately fast water, a white wolf showed her head from behind a willow on the bank above us. Soon she came out in the open to stare at us. There were some spruce trees on the bank and a big snow patch on the hill behind her. She seemed curious, and watched us without apparent alarm. Fred's camera was always in a waterproof pack immediately ahead of him; a hard pull on a zipper and it was in his hands. Being ready was only half the battle in a moving, twisting, bobbing canoe, but a remarkable memory was captured. The wolf's coat was tidy and neat and almost pure white, and her nose deep black. She was lean, with very narrow hips, narrow shoulders, and long legs, and she looked young and healthy. Perhaps she was immature, full of curiosity, and not yet as cautious as might be wise. Even Alex, who has seen many wolves, was thrilled. There is something deep in us that responds to the sight of a wolf, something primitive. No other single animal is more emblematic of the wild North. Wolves are surrounded in legend and fear, and even loathing, but in truth they are social animals, devoted to family and kin. Attacks on humans are rare.[2]

Camp was set up near three wolf dens known to Alex from past trips, about ten miles from where we saw the wolf. Massive sand dunes towered above us. Small heath shrubs were ablaze in the bright colors of early summer bloom: Lapland rosebay with pink to purple rose-colored flowers, and Arctic azaleas with tiny pink flowers. Fragile-looking mountain avens with large white flowers and yellow centers were blooming in patches, adapted so that their flowers rotated to face the twenty-four hours of sun at all times. We set up our tent in a field of Arctic azaleas.

A "little" walk before dinner to scout the wolf dens turned out to be much longer, in this case about five miles before it was over, but it was worth it. Along the way we saw abundant old moose, musk-oxen, and wolf prints, and "not too old" huge grizzly prints. Less than twenty-four-hour-old grizzly prints were found at our tent site. The first two wolf dens were empty, although disappointment was tempered by seeing great fields of Lapland rosebay in bloom on the sand.

FIGURE 21 Arctic summer. Lapland rosebay.

After a brief discussion ("Do we want to go on?"), the group tramped single file across a huge dune, and eventually came upon fresh, large wolf prints and scat. It was obvious that we were close to an active den. Cumulus clouds were gathering above us, casting shades of gray on the scene with patches of bright sun. Alex motioned to us to be quiet, and to approach cautiously. We crept forward, bending low, anxious and with hearts pounding. A sleeping black wolf just outside the den departed without a peep. Soon we heard wolf barks and woofs, and two large adult white wolves emerged from the den behind the bushes, thirty feet away at most. One wolf, undoubtedly the alpha female as she was first out, came around a shrub into the clear and stared at Fred for a moment at about fifteen feet. Both of them froze for just an instant. There was no chance for a picture before she and the other wolf bolted for the heights of the dune behind us. Standing proudly high above us, she repeatedly looked behind her as though to keep the other wolf out of danger. Arching her head up, she howled and woofed at us, and then moved her position a bit before resuming howling. She had pups in the den, and was most unhappy that we were there. We finally left after about a half hour. Alex glimpsed two pups in the den, but did not reach in to pull them out, as he once did only to find an adult wolf was still in the den. We were exhilarated, and did not feel tired after walking about a mile back to camp over streamside boulders. Our heads were buzzing from all the amazing animals we saw. It was very bright at 9:30 when we retired, but there was no trouble sleeping.

In the morning we paddled downriver a short distance and walked in to the wolf den we visited the evening before. Disturbed wolves often move their pups to a new den, but we were serenaded again by the howling white mother from the top of the dunes. She was up there before we arrived, and may have kept a vigil during the night, undoubtedly aware of our campsite. She was the prototypical mother protecting her family, magnificent, apparently unafraid but certainly upset that we were there again. Joyce felt akin to her and urged us to leave the wolf's domain. When we left to paddle downstream, the wolf followed us along the top of the dune, clearly silhouetted against the sky, head arched upward and howling for many minutes before we finally lost sight and sound of her. None of us will forget her. She provided an intimate and strong if brief connection with the wild, giving us chills and moving us deeply. We were glad the programs to poison wolves were abandoned many years ago, and that hunters are not allowed to "sport"-shoot wolves from planes, as some do in Alaska.

Whitewater dominated much of the next two days, some easy, other times more exciting with two- to three-foot haystacks, which elicited many shouts and whoopees. We shipped cold water in the bow, adding excitement to Joyce's day. Alex told us that one canoe tipped over in this stretch last year. While we were going downstream, he asked where we did our canoeing. We reminded him the last real trip was twenty-five years ago. He was silent for a moment and then said, "I guess it is like a bicycle: once you learn, you can always do it." Walking on river's edge, we guided the canoes by five rapids, wet to our thighs and cold most of the day before camping on a treed oasis to get out of the wind. July 1 arrived, and certainly it was beastly hot back in North Carolina, but here it was cold, and there was ice on the hills. The river was picturesque, with slate and sandstone cliffs, big sand dunes and eskers. The sky was huge and deep blue, with white and gray clouds rolling over us. There was not a scrap of civilization, only a pair of high jet contrails to remind us we lived in a modern world. Winds and whitecaps frequently made it hard work. We were warm when we were paddling, but cold when we were not.

Kitty and David paddled well together, always in good rhythm. Stu and Alex frequently were out of rhythm; Alex paddled powerfully with his almost sixty-strokes-a-minute, short, hard, "Canadian" stroke technique, while Stu paddled as did we with a longer and slower stroke. We wondered whether Stu paddled bow with Alex on each of his previous twenty-six trips with Alex.

Cliffs and dunes gave way to flatter land. At noon of the sixth day we intersected the Thelon and left the secret river for good. As we approached the Thelon, the river seemed to rise to greet us. Joyce in particular was astonished by the deafening noise, and the sight of spew and foam celebrating the joining of the waters. Intense wild forces were at work here. The sound and sight of the great river were both exhilarating and intimidating. The Thelon is a powerful river, coursing about seven hundred miles north from the tree line to Baker Lake. Along the way, it cuts through the heart of the central Barren Grounds, and the center of the Thelon Wildlife Sanctuary. There is a well-known oasis of spruce trees along its borders, well above the tree line. The Inuit once traveled here to collect precious wood for kayaks and other purposes. We had seen no Inuit, Thule, or Dorset camps along the way, and there were none until farther down the Thelon, in the Beverly Lake area. It was nearly four hundred miles downstream from here to Baker Lake.

Huge rapids barred the way, much bigger than anything we had encountered. There were large boulders and three- to four-feet-high chunks of ice along the shore, left from breakup a few weeks ago. Tall sandstone cliffs pinched the river, and there were haystacks, holes, submerged rocks, whirlpools, eddies, and generally scary water for at least several hundred yards. After scouting the rapids for some time, and telling us he usually paddles all the canoes through these rapids himself, Alex decided we could do it. He cautioned that we must stay out of the big waves and must avoid being swept into the cliff face downriver, either of which would probably flip the canoe.

"If you go in, it will be half a mile before we can fish you out, and you could die of cold. Stay to the left after you pass the nose on the cliff, and avoid that big wave, then go hard to the right and get out into the waves to avoid the cliff face on the left. Do not hit the cliff."

These were probably only Class III rapids down South, but if Class IV is defined as a rapid that can kill you, this was Class IV.

"Glad we did it.... Glad we did it," said Joyce when we pulled up to rest.

Just below, another river entered with a blast of powerful water. Before long we camped on a nice hillside spot. There were two little ponds with ice piled up on the shore, and Joyce bathed in one. Fred watched Joyce's face immediately contort with the cold; she took a very short bath. Later she compared the feeling to a dull but deep pain. Apparently this bath satisfied the cleanliness gods, since she was not to try this sort of bravery again on this trip. Kitty went in also, and showed off by staying in the water for a couple of minutes. It took her another half hour in her sleeping bag to get her core temperature up again. Fred elected to stay dry and dirty. All napped before dinner. A massive double rainbow appeared over the river, falling down from the skies onto the open tundra. Later another rainbow arose while we fished the flat waters by camp. We were warned that once on the tundra and in the canoe, we would never want it to end, and would want to keep coming back. "Tundra fever," they call it, a longing for wildness, and solitude, and satisfaction in a day's hard work. We felt healthy, relaxed, fulfilled, and competent.

This section of the Thelon is full of small spruces up to thirty feet tall, in a narrow band no more than one hundred yards wide along the river edge. The oasis runs for many miles, but further downstream the spruce come to an abrupt end. Most trees were scrawny, but some were three feet in diameter in favored places. We had seen scattered small spruce along the secret river, but here on the Thelon oasis they were quite abundant. The trees undoubtedly were older than they looked; Ernest Thompson Seton counted annual growth rings on a similar small spruce not far from here during his epic 1907 paddle into the Barrens, and found that a tree only eight feet tall was over three hundred years old.[5]

It warmed up. Paddling was easy, but mosquitoes appeared. The flowers were really out now, at least the early bloomers. There were many flaked points, stone tools, and scrapers at Grassy Island, and a better collection at Cosmos Point, named after the Russian satellite that came down near here twenty-five years before. Alex kept a cache of museum-quality arrowheads at the latter site, buried under sand and rocks. Old caribou crossings also were evident in many sites, explaining the arrowheads. We were getting good at predicting when the skies would light up with gold and orange fire; setting an alarm allowed us to see a grand sunset at 11 p.m. and a sunrise at 2 a.m. reflecting on the black river below camp.

Wolf and grizzly tracks were seen within two hundred yards of one campsite. We checked out a wolf den that had been active for sixteen of the last nineteen years, and Alex was crestfallen when we discovered it had been destroyed by grizzlies. Head down, shoulders slumped, saying little, he looked as though he experienced a death in his family; he counted on seeing wolves here, as he had done for so many years. Although there were fresh signs of grizzly (mother and two cubs) and musk-oxen, there were no fresh wolf signs. Some compensation was provided by finding a gyrfalcon nest in a spruce tree, with at least three white chicks in the nest and all white parents circling overhead. The white phase apparently is rare. These birds are heavily protected because they are worth a small fortune to rich individuals who use falcons in sport hunting. We saw tundra swans, common loons, yellow-billed loons, a marsh hawk, black scoters in large numbers, two bald eagles, mergansers, and various geese and ducks. A snipe engaged in its courtship dance overhead.

The bear alarm was on every night, but we saw no bears until the very end, which was fine by us. A quite blonde mother grizzly and her two-year-old cub appeared through the willows near shore. She stood up, charged a short way into the water and woofed at us. Frankly, she was scary. Fred whispered to Joyce to paddle closer to enable better photos, but she sat perfectly still, allowing later that she had misunderstood the message. The bear's bluff succeeded: we did not move closer. It undoubtedly was just as well that we stayed where we did. A solitary old moose with half-grown rack lay down in the willows as though to hide from us, but he was huge and there was no hiding.

Hornby's cabin was not far from the river but was surrounded by dwarf spruce, and finding it required knowing the country. Hornby was a crazy and fearless little man of five feet and one hundred pounds, and a romantic fool who died of starvation here along with his nephew and friend in 1927. They were the last people to try to live here. They left a diary behind in a potbellied stove, which helped to increase their fame. In tragedy, Hornby became a kind of heroic figure.[6]

The water was crystal clear, even after rains, and we drank from it without fear of germs. We were very far in every way from the ugly algae blooms that sometimes marred the river below our house in North Carolina during summer heat and drought. The aches and pains of the first days had gone entirely. Extending toilet paper by use of spongy green moss was a new skill. We were careful to use our new talent as far from camp as possible and to cover up our "business" so others did not inadvertently come across it. Stu said we had lost weight, but we did not know how that was possible with all the food including chocolate bars we had eaten. Alex said he typically loses a pound a day for the first twenty days, and cited clients who lost thirty pounds or more and about twenty years in appearance.

"I should call this a beauty camp," he said.

Alex regaled us after a fashion with bug stories from earlier trips: "There are always mosquitoes an inch thick in the soup pot.... They die after flying through the steam, and fall in ... They eat us and we eat them.... The flies not only cover us with a layer of black but they are in multiple layers, and crunch when you flex your arm."

We felt lucky to have had such a cold year. We drank the last bit of scotch sitting among a field of mountain avens on the bank looking at the river.

Alex offered us a place on his by-invitation-only trip to the Back River in 2006, two years hence. The river is full of challenging rapids, which is why Alex generally restricts it to people with whom he has paddled. His caution was for reasons of ensuring a modicum of technical competence as well as emotional stability among his fellow paddlers, and their willingness to keep going when fatigued.

Stu was amazed. "Do you realize what he just did? He offered you the graduate course."

We were flattered and with only one brief look at the other agreed to go. The region abutting the Back River was near the former territory of the Utkuhiksalingmiut, who impressed us so a year earlier in Baker Lake.

The plane for leaving the Thelon arrived early, guided by a satellite phone call from Alex who sent our exact GPS location. The canoes were disassembled so they could be stacked inside each other and tied to the plane's floats, to be dropped upriver. Alex would pick them up at the start of his next trip. Many lakes were still covered in ice. We knew we were fortunate to see and experience one of the last great stretches of wilderness not only in North America but in the world.

After we returned home, Stu sent us a reprint of an article he published in 1991 in the journal *Musk-Ox*, regarding the starvation deaths of sixteen Utkuhiksalingmiut at Garry Lake in 1958.[7] Besides knowing Doug Wilkinson, the Inuktitut-speaking RCMP officer who comforted Kikik at her trial, Stu had traveled by dog team to Fury and Hecla Strait, and fished through the ice in deep winter with Inuit friends. During our time together on the secret river and the Thelon, we heard none of this, and failed to appreciate the depths of his understanding of the Inuit and their history. We traveled together yet were in separate canoes and separate tents, and were tired at day's end. We did not connect to each other as we might have. The angst engendered by this realization increased beyond measure when we heard from Alex that Stu was hit and badly injured by a truck while he was riding his bike in the mountains. Stu suffered for weeks in the hospital. A leg had to be amputated. Eventually his heart gave out. Before he died his sister wrote a sensitive note saying how much Stu enjoyed the photos of the trip that we sent him, with musk-oxen, caribou, wolves, gyrfalcons, Arctic wildflowers, whitewater, and beautiful eskers and ice. After his death she wrote again, saying how the photos sustained him, reminding him of all the good years he had paddling with Alex. As a tribute, Alex established a new set of photos on his Web site, along with text eulogizing Stu. Our photos are on that site, part of Stuart's memory, and we imagine they are being broadcast to the ether where Stu can still see them. Stu was a Presbyterian minister's son and undoubtedly is in heaven. We trust the photos help his spirit recall being on the Barren Grounds in unspoiled wilderness, in God's land.

7

GOLDEN CARIBOU:
BAKER LAKE AND THE KAZAN RIVER

Inuit ku, the river of men,
Highway into the heart of the sky;
Caribou tracks on my soul.

J.K.[1]

JULY–AUGUST 2005 The Kazan is a famous and very large river, connecting a series of massive lakes. The lakes remain ice-bound well into July, and some never thaw completely. The river flows directly through lands occupied by the Caribou Inuit within our lifetimes, and some call it "Inuit Ku" (River of Men).[2] By flying into Baker Lake, and then taking a bush plane upriver, we could paddle down to Baker Lake, starting and ending in the same place. This would enable a substantial visit with our Inuit friends again. Encouraged by the trip down Alex Hall's secret river and the Thelon the year before, we thought we could canoe it even though portaging would be inescapable. After much discussion in the confines of our warm and comfortable home, we decided that our souls needed this trip.

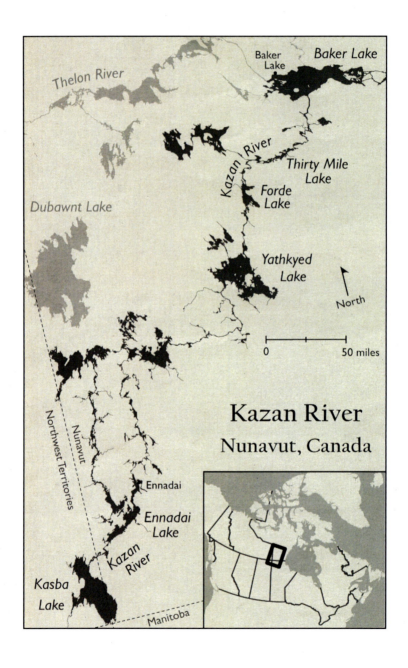

The river flows from the border of Saskatchewan and Manitoba north to Baker Lake, traversing some 530 miles. At its southern regions, there are small trees, which are replaced by thickets of willows and dwarf birch and open tundra and bare rock as the river flows north. The Kazan was the home for several Inuit communities in the nineteenth and early twentieth centuries, including the Ahiarmiut, Paallirmiut, and Harvaqtuurmiut, who were drawn to its shores by the Qamanirjuaq caribou herd. Intermittent starvations eventually forced the Inuit off the land, and it has been fifty years since Inuit made their homes on the Kazan. Nevertheless, there is a rich archaeological record of their former lives here, and elders in the community of Baker Lake still remember life as it was in the old days. In 1990 the Kazan was designated a Canadian Heritage River in recognition of its importance to the Inuit.

The first European to paddle the Kazan without a guide was J. B. Tyrell, who led a party down the river in 1894. To this day the number of parties that have canoed the river probably amounts to only a few hundred.[3] The area is isolated and presents many challenges. The river can be canoed only from early July because of the ice. Onset of winter cold in September limits travel to the middle or end of August for all but the foolhardy. We know a member of a party that went down the nearby Dubawnt River in September, got trapped in the ice, and suffered the death of the trip leader after a canoe spilled.[4] Most trips start near the river's origins, because the headwaters are accessible by road, and cover the entirety of the river, often taking forty-five to fifty days to make the journey. These are epic trips for most people, done only after years of planning.[5]

Desire is easy. How would we do it? And could we do it, really? Canoeing this river from start to end was not practical for many reasons. Only one guiding outfit included the Kazan in its repertoire. Wilderness Spirit was willing to make this trip with only three of us. We would do the lower 170 miles, starting at Yathkyed Lake. Allowing time for wind delays, about fifteen days would be needed to reach the takeout point at the mouth of the river at Baker Lake. One major portage would be required to get around Kazan Falls. Fred's brother Rand had a lot of kayaking experience, had committed to the 2006 Back River trip with us, and was willing to take on the challenges of this trip. Alex Hall cautioned us about the dangers of the winds on the big lakes, and kindly sent an article about the archaeology of the lower Kazan.[5] Canoeing friends shared their maps and previous trip notes, and added their cautionary tales about winds on big water.

The prospect of portaging 70-pound packs and canoes was worrisome, even if such weight was a mere trifle compared to the 150 to 250 pounds that voyageurs, early explorers, and all the aboriginal people of the North once routinely carried on long treks and portages. It was imperative that we be as well prepared as possible. We became members of the community who meet regularly in the gym to sweat and groan either to look better or to be healthier, or in some cases to take long hikes or long paddles into the wilderness. For five months we worked out on the treadmills and StairMasters, with stationary weights and on various muscle-building machines. Sometimes Joyce pushed Fred to try harder, and sometimes Fred goaded Joyce to stay with it. In June we left for Westport for a few weeks of canoeing and bike riding. Rand and his supportive and enthusiastic wife Adrienne joined us for a week of paddling on the Westport River, sometimes in rain and wind. Rand was hitting the gym hard, and had stayed in pretty good shape throughout his life. He was five years younger than we at sixty-four. Back home in North Carolina, preparations continued until finally we packed our gear and left for the North. Hopefully our physical condition was good enough for canoeing the River of Men.

We flew to Winnipeg on July 19, and then to Baker Lake via Calm Air, where we met our guide, Rob Currie, for the first time. He was forty, tall, rugged, and handsome (so Joyce revealed), and a very experienced canoeist, although only once had he canoed the Kazan, and that was nineteen years before. Rob quickly put us at ease, and soon we learned of his quiet humor, culinary abilities, and confident guiding skills.

Joyce contacted Hanna, and she was at the airport to welcome us along with her attractive daughter Patricia. We all convened at Baker Lake Lodge for tea and talk, and to the apparent shock of the owners who seemed surprised we had Inuit friends. Since our fortieth-anniversary trip to Baker Lake in 2003, we spoke with Hanna by phone on several occasions. Initially the calls related to the status of the first wall hanging her mother Ruth did for us, or to a second work on cloth Ruth did in 2004. Later calls were to pass the news. A woman disappeared and died in a blizzard not far from town. Another winter storm trapped two people, one of whom froze his leg and

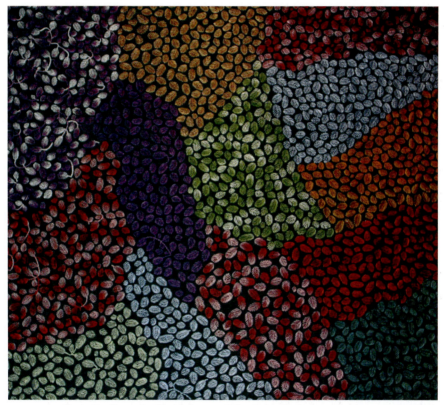

FIGURE 22 Ruth Qaulluaryuk, *Untitled*, 1994, Baker Lake. Bird appears to be emerging from tundra background. Entire image consists of detailed feather stitching with embroidery floss on a wool stroud base.

required amputation. The other was spared by wearing skin clothes, which are much warmer. Hanna told us she walked to Grandmother Oonark's grave, after Ruth and Josiah admonished her to walk more. "They say I don't appreciate the tundra the way they do, and I need to get out there." We also wrote, not only to Hanna but also to Ruth and Josiah, knowing Hanna would translate our letters for them. Sadly, many of the calls to Hanna in the past year related to the illness and death from cancer of her father Josiah, who died only a month before we arrived.

Before we departed for the river we were able to visit Ruth in her home. Ruth appeared smaller in size and withdrawn, but brightened noticeably with a gift of embroidery floss. In addition to the death of Josiah, her sister-in-law, Janet Kigusiak, also died recently, so there was abundant reason for her depressed affect. We talked briefly of Josiah's passing. Hanna was glad that her dad was no longer suffering. She spoke of how he always carved shamans, but he read his Bible twice every day, and now he was at peace. We left a framed picture of Josiah with Ruth, who stared at it silently without twitching a muscle. A son thanked us for the frame. The home seemed empty without Josiah's jovial laughter and jokes, which we remembered very well from our visit two years prior.

There was time for another visit with Miriam Qiyuk. After showing a picture of a wall hanging of hers that was in our collection, she commented that it was her favorite piece, her best work. She wished that she had kept it, but sold it because she was so poor. Like most of her work, this work on cloth is a pastoral scene of an Inuk woman's world, replete with gentle images of fishing, animals, and birds, surrounded by colorful owls whose meaning is unclear. Later, her son Paul also showed us another of her wall hangings, depicting a sad young woman being carried away by a man against her wishes on a qamutiik. Lest there be any doubt, her maiden name was inscribed below in syllabics with an arrow pointing to her, and her husband Silas's name was similarly inscribed under his image. Decades later, she apparently was still angry at how she was abducted, once a common way of securing a wife.

FIGURE 23 Miriam Qiyuk, *Untitled*, 2005, Baker Lake. Detail of wall hanging showing her abduction as a young woman by her husband-to-be.

FIGURE 24 Miriam Qiyuk, *Untitled*, 2004, Baker Lake. Wall hanging depicting a woman ice fishing, surrounded by animals.

But it was time to prepare for our canoe trip. Rob gave us blue barrels in preference to canvas packs because they were absolutely waterproof. With a harness, they could be carried easily. They also floated and were relatively impervious to bears, which did not totally reassure us. Wilderness Spirit bought new foldable and relatively light Pak canoes for this trip, their first experience with these craft. The only available plane had tundra tires rather than floats, and thus foldable canoes that could be carried inside the plane were necessary.

Another group was at the lodge, a well-known Inuit art collector and his friends, who had flown in the most expensive Burgundy wines for their dinner table. They planned on using the turbocharged single-engine de Havilland Otter to fly to a lodge north and east of here early the next morning, exactly when we were scheduled to fly out. It was not surprising when we were told at dinner that we would fly in an hour.

At 7:30 p.m. on a cool and slightly rainy evening we were taken to the airport, where we loaded our gear and took off, leaping up into the mists. Pilot Tim knew a good spot for a landing at the upper (south) end of Yathkyed, three miles from the water. Landing at the south end, however, would add another forty miles of open water to our trip. The thought of three miles of unplanned portage over swampy terrain with packs up to seventy pounds fully loaded almost caused a heart attack.

Flying low over very wet and totally treeless tundra, we saw white wolves and a herd of musk-oxen materializing out of the gray mists below us. Lakes, ponds, and rivers dotted the landscape, but there was no ice. The Kazan looked very big and strong, with multiple rapids. Most of the land was swampy and rocky, without hope for landing the plane. After circling over the north end of Yathkyed several times, we finally found a gravel bar about a mile from the water. Tim first tested the firmness of this little strip by touching without landing, and then came around and landed in about two hundred feet of semisoft gravel. He wheeled the plane around in a tight half circle, at the edge of a drop-off, leaving deep ruts in the gravel but in position for takeoff. A united sigh of relief was felt more than heard.

We unloaded quickly. Tim took off as soon as possible, jumping up into darkening skies with only a shaft of yellow light streaming through a break in the clouds, quickly becoming a speck. Alone with our thoughts and unspoken anxieties, we felt lonelier than last year when we were left on the secret river. The size of the group may have made a difference; with only four of us we were the smallest safe size for this trip. This country seemed less hospitable, perhaps because we were on an elevation and could see far across the Barren Grounds and across a vast lake. It was cool, and a soft light spread over the lake. After pitching camp and having some tea, we retired to our tents and crawled into our bags.

Rob prepared a delicious breakfast of real eggs, onions, and bacon—a great bear attractant, but we were about to leave. Assembling the Pak canoes took over an hour. Each canoe was seventeen feet, and was supported by aluminum tubular struts inserted into heavy polypropylene sides, with a neoprene bottom. Carrying our gear to water more than a mile away would tell us whether the work in the gym was sufficient. Traipsing over the muskeg and swamp was hard labor, requiring perhaps three hours. At least part of it was downhill. The Labrador tea smelled great when we stepped on it, but pleasant odors were hard to appreciate under the stress of the load, and the attacks of the mosquitoes.

We were in the extreme northwest corner of Lake Yathkyed, which means "white partridge" or "white swan" in the Dene language. The Inuit call it Hikoliqjuaq, which means "great ice-filled one" in Inuktitut. The dual names reflect its shared membership in two cultures, but inhabited at different times. It is a huge lake, only slightly smaller than Baker Lake. David Pelly and his Raleigh Expedition team of thirty-two young people from around the globe paddled through here in 1988, and they were thrilled to see twenty thousand caribou on the northeast shore.[6] In 1922 Rasmussen encountered the Padlermiut (Paallirmiut) somewhere close to here.[7] He had long conversations with the head man and eminent shaman Iqjugarjuk, from whom he learned much of the beliefs and customs of the people. A recent famine had passed, and together they feasted on caribou. Dessert of caribou fly larvae was less appetizing, but Rasmussen recorded that they "gave a nasty little crunch under the teeth."[8] Most distressing was his discovery that Iqjugarjuk was a more modern man than he suspected when Enrico Caruso's voice suddenly blared forth from a gramophone in his tent. In addition, inside the shaman's tent were tins and metal cooking utensils and other items garnered from trading posts in exchange for fox skins. Rasmussen concluded that he came a hundred years too late, although in other respects the people he found near here were still living typical hunter-gatherer lives. Eighty years later, most of the world was very different, but here it was unchanged except for the absence of Inuit.

No sounds of Caruso remained, only silence, as our group of four set out paddling onto the big lake. The first moments on the water were special, a memory to be savored. Sitting in fully loaded canoes, we were inches from the water, a very intimate connection. It was almost as though we were immersed in the water. We dug in and paddled hard, quickly finding our rhythm, following the shore. After no more than six miles, with some headwinds and whitecaps, we entered the Kazan's quite fast water, thankful that we did not have to start from the distant south end of the lake. We scouted a Class III rapid and ran an easy slot on the side. One possible campsite was interesting, because it had lots of tent rings, caribou tracks, and hair, but we paddled on to another site that also had many caribou tracks and an inuksuk. Mosquitoes were in evidence, but a big green bug tent for cooking and eating that Rob always carried made all the difference.

Two wolves appeared on the other side of the river after dinner, one small and white, the other very large and gray. Two cow caribou they were stalking saw them and departed over the ridge. The white wolf saw us and started to swim the river. We watched with binoculars, fascinated, hoping she would come, but anxious, too. Fear of wolves is not easily put aside, even though we knew their hunting instincts are tuned for caribou, not humans. Alex had seen wolves amble into the midst of camp, just to see what is going on.[9] The river was swift and wide, and it may have seemed too much effort merely to satisfy her curiosity. We were disappointed when she turned back.

The next couple of days were not difficult. The weather was cold and gray, with some rain. Clouds sometimes hung low over our heads, a roof over the land. The river edge was lined with boulders left by the melting ice of spring, and sometimes was so wide it seemed like a lake. In places the water narrowed, and there were relatively easy but turbulent rapids. We shipped water at times. Paddling a river with good flow like this should be easy, but wind happens, lots of wind, usually from dead ahead or the sides, which is worse. South breezes at our backs were wonderful while they lasted, which was not long. Flat, calm water was welcome, except it often meant the black flies were out in full force. Without a good breeze they were troublesome, and head nets or bug shirts were mandatory. Joyce seemed more sensitive to fly bites and for many days had swollen eyes and facial edema from the bugs, unbeknown to her.

Occasional caribou and musk-ox were seen at a distance, and at times we encountered clumps of caribou hair. Sandhill cranes were not far away on the tundra at potty breaks, and we encountered an old Inuit hunting blind. A low fence of modest-sized stones was arranged in a long, straight line, apparently to channel caribou during the hunt in the old bow-and-arrow days. Ptarmigans, all with chicks, jumped up from hiding several times while we were taking breaks on the land. The birds were very pretty with white under their wings and red around their heads and necks. In winter they would be white.

The wind switched hard off the port bow or directly in our faces as we entered the south end of the huge, twenty-five-mile-long and fifteen-mile-wide Forde Lake. Waves crested at our gunwales, threatening to swamp the canoes. Fatigue set in alarmingly, and at noon we pulled in to get behind the shelter of the only visible large rock. Everything was wide open here and flat. The winds continued unabated, and we made camp early. We committed to rise at five the next morning to get ahead of the wind. Otherwise we were going to need to paddle at night.

Luck was with us, as the breezes dissipated. Hard paddling up Forde Lake under very calm conditions resulted in a gain of over twenty miles before lunch. We were totally engrossed in the effort of paddling, staying not too far off shore lest the wind catch us. There was a sense of being on an endless open body of water, where progress was slow, and we inched forward to a distant point of land. It seemed we were suspended in time, in a sort of endless cocoon where nothing else mattered. We were in good rhythm, and glided. A simply spoken "hut" by either one of us, and we switched sides for paddling. Little was said. Talk can be a distraction; silence was comfortable. Our private thoughts drifted from the Inuit who used to live here to the look of the clouds in the skies. There was no thought of the small or big problems of everyday life; the stock markets did not exist, and world politics were totally irrelevant. Other than a satellite phone for emergencies, we were utterly out of contact with the electronic world. Nothing existed but the immediate present.

Eventually the wind did come up. It always does. The packs and our bodies acted like sails, and the canoes did not track especially well. In the early afternoon, near the north end of Forde Lake, we elected to camp. It felt good to stretch out on the sleeping mats on the open tundra under a warming sun. Lush patches of peaty, moist tundra nearby were covered in cloudberries, and their sepals were a lovely rose-pink color. We could barely see the south shore despite the clear air. Many loons, Arctic terns, and one bald eagle were on the water or flew nearby. A twenty-seven-inch lake trout supplemented an excellent dinner with Rob's excellent dried caribou obtained from an Inuk friend. Exhausted, we started to retire at 8 p.m. when two bull caribou swam by in the soft light. They veered off at the last minute and climbed out thirty yards down the shore, shook off the water, and started to eat lichens immediately, oblivious to us. Soft new felt covered their antlers, glowing in the backlight of evening.

The immediate challenge was to paddle directly across the five-mile breadth of upper Forde Lake. Out in the middle there would be no easy escape if Sila, the Inuit spirit power of the weather, decided to blow up the

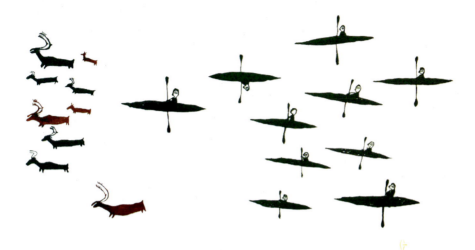

FIGURE 25 Luke Anguhadluq, *Kayak and Caribou*, A/P stone-cut print, 1971, Baker Lake. Reminds us of the caribou crossings on Forde Lake and the Kazan.

FIGURE 26 Caribou swimming Forde Lake at dawn.

winds all of a sudden. We were on the water before six in the morning under beautiful, calm conditions, with low clouds overhead and a bit of golden light peeking through on the edge of the land across the lake. Before we had gone more than one hundred yards, we saw another pair of bull caribou swimming across the wide lake, with a third trailing them at some distance, coming right to us in the soft pastels of early morning, an amazing spectacle. We paddled right behind them, and followed closely without seeming to disturb them. The Baker Lake elder's accounts of life on the land in the old days[10] stated this was the way to approach them by kayak, only pulling up alongside to spear them when they neared shore. Strong swimmers with their broad hoofs adapted both for walking on snow and paddling through water, they left a small whirlpool just behind their tails. They appeared vulnerable as they clambered awkwardly out onto shore. Shaking as soon as they stepped onto land, their great, broad antlers were immersed briefly in a dense, soft halo of spray. We recalled our print *Kayaks and Caribou* by Luke Anguhadluq, which depicted in very simple yet elegant lines the excitement of the chase, with one kayaker ending upside down.

The Inuit name for the river from this point north is Harvaqtuuq (the big drift); the people who lived, hunted, and fished there were the Harvaqtuurmiut, or people of the big drift. The Harvaqtuurmiut had a song about hunting caribou from kayaks that was particularly apt for us old-timers attempting to recapture their youth:

> *Happiest am I*
> *In my memories of hunting in kayak.*
> *On land, I was never of great renown*
> *Among the herds of caribou.*
> *And an old man, seeking strength in his youth*
> *Loves most to think of the deeds*
> *Whereby he gained renown.*
> ULIVAK'S SONG[11]

A huge caddis hatch emerged on Forde Lake, and trout rose all over, as far as we could see, with little dimples everywhere as a trout sucked in a caddis fly. On the far shore, dead caddis flies were stacked in piles an inch deep. Fearful of the winds, we did not set up the fly rods, but just kept paddling. With no wind the black flies were after us. People gasp and utter sounds of disgust when they see our photos of Rand surrounded by a cloud of black flies over his bug-jacket-covered head, but as Alex said, the flies keep the meek off the tundra, preserving the peace for those willing to put up with them. It wasn't really that bad once we got accustomed to them and accepted them as part of the experience. A breeze got rid of them. A massive gaggle of geese on the water rose as one as we approached, amazing us with a symphony of goose calls. The skies over the far side of the lake were a blue-gray color that reminded us very much of the icebergs off Ilulissat. What a sublime paddle, in a vast and unforgiving country.

FIGURE 27 A modest cluster of black flies has Rand trapped in his bug shirt, on the Kazan.

Finding the exit from Forde Lake into the river was difficult, even with the help of Rand's and Rob's GPS receivers. Everything on shore looked the same. Only when we were extremely close did we recognize the river. Lunch on a hill with a large inuksuk at the top was a welcome chance for rest, and while inspecting the inuksuk we saw a small herd of cow caribou and their calves passing just below us. A herd of musk-oxen ambled slowly in the far distance. Each inuksuk was different, and we wished we could understand their unique meaning. They often were asymmetrical and pointed directionally, indicating a place to hunt or fish, or a direction to travel. They might point to a place for ice fishing: a location close to the water meant that fish were close to shore; a location distant from water meant it was necessary to go out toward the middle to make a hole through the ice.[12]

A big blow buffeted us almost immediately after we got back into the canoes, but it did not last long. We were paddling through the heart of formerly densely occupied Inuit sites, and made camp at a site with many tent rings near water's edge. Fall migration camps were near the water, whereas spring migration camps were on higher ground where the snows melted first. Tracks indicated we were at a major caribou crossing. Joyce caught three nice lake trout, about six to fifteen pounds, and we had orange trout fillets for dinner. Rand named Joyce "Kazanuuti," his term for the great fisherwoman of the Kazan.

After dinner, Kazanuuti suddenly declared, "Men, I may be crazy, but there are many caribou in the water." All eyes turned and focused; although the wind had died down and the black flies appeared, we sat mesmerized. Caribou cows and calves were swimming across the river above the rapids. The water was high and fast, and some got swept downriver. The caribou often returned to make the same mistake again before learning to enter farther upstream. One group apparently lost a calf in the rapids.

That night Fred noted a small boil on the extensor surface of the first finger on his right hand, and started to take dicloxacillin for presumed staphylococcal infection. Kazanuuti was feeling flush with confidence and made a few unnecessary but humorous (to her at least) comments about the precision of the diagnosis. The boil awakened Fred in the night. It grew rapidly and began to throb, turning gray with internal hemorrhage and looking necrotic. Fred lanced it with a scalpel, which released copious pus. A regimen of saline soaks commenced, which caused frequent delays for the remainder of the trip.

The weather was alternately cold and gray with whistling wind and whitecaps, soon changing to warm and sunny, and then to rain. We paddled through multiple rapids, one of which was Class III with scary ledges and holes. Three sturdy inuksuit and many Inuit tent sites were found above the rapids. At lunch we repaired Rob's and Rand's canoe, which was punctured. Our canoe was repaired the day before, after we bumped a sharp rock on the side. A bald eagle, a few caribou, and three musk-oxen appeared at a considerable distance. Above the roar of another rapid, we camped in a boggy site, but were saved from bugs by a blessed wind. The rapids looked easy, but experience taught us that they looked a heck of a lot bigger in a canoe.

After a night's rest, we enjoyed paddling on a beautiful, warm, utterly calm day, entering the upstream end of Thirty Mile Lake. The lake is narrow but long, and contains many islands with inuksuit and some old gravesites.[5] On one island we found a very large collection of split and fractured caribou long bones, from which the marrow had been harvested. There were stone tent rings and old stone caches for storing meat, and pieces of wooden kayaks. An old copper knife blade was testimony to trade with the Copper Inuit to the northwest. A caribou "fishing rod" once was used with sinew lines to fish through the ice. Low kayak stands made of stone remained. Everything was left exactly as we found it.

A tundra grizzly stood up on the low tundra hills to get a better look at us, paddling close to shore in our little red canoes. He retreated in stages up a ridge, repeatedly pausing to stand and stare at us. He was as aware of us as we were of him. We paddled only another two and a half miles before we made camp on a low, sandy beach at a narrow spot on the lake, designated "Panningunianaag" on our maps. The site seemed a bit close to where we saw the bear. Refuse was left here by geologists who drilled a bunch of deep core samples of bedrock, and left them in trays along with other scraps of camp stuff. At no other place did we see any trace of manmade detritus. There was a ridge well behind our tents, which Rand and Rob climbed, where they found a vast collection of inuksuit, tent rings, a stone quarry, and other evidence of habitation by as many as one hundred people.

Fred's hand did not improve much and appeared nasty. A second lesion developed on the back of the hand, which he also lanced. The whole hand was swollen, and the primary lesion was gray-black and looked dead. (The finger squirted green pus in two directions from underneath the dressing when stress was put on it.) Antibiotics clearly were not working. We discussed the risks and possible solutions, including medical evacuation. If the hand worsened, it might be impossible to paddle and dangerous to life and limb. We elected to rest a day and concentrate on soaking the hand. Fortunately, the hand slowly improved. Three months later, back home again, the infection recurred. The cause was a resistant strain of staphylococci that has become a serious problem in otherwise healthy people.[13]

When relieved of nursing duty, Joyce floated a dry fly in a tiny rift of current and pulled forth a fourteen-inch fish; the large dorsal fin indicated the fish was a grayling, her first ever. Meanwhile, a pair of sandhill cranes and their chick sauntered along the shore until they spotted us in the bug tent. A wonderful golden sunset closed the day gloriously. The days were shorter now. The next day was quite cold, windy, rainy, and miserable. The weather was in charge here. We did a lot of reading in our soggy little tent, and took naps.

In early evening we climbed the distant ridge to inspect the inuksuit, graves, and tent rings that Rob and Rand found the day before. Marveling at a pair of tall and dramatic inuksuit, each standing on two legs and at least as tall as a man, we heard a loud yell from Rob far below. Using Joyce's small binoculars, we saw the grizzly approaching camp. Rob always carried a shotgun, and was marching toward the bear, Rand following closely. Sounds of a shot boomed up the ridge, followed after a few minutes by two more.

The last warning shot finally scared the bear, who retreated toward us. Seeing what was unfolding, we said nothing, but started walking the long way around, avoiding the ridge lest the bear depart along that path. Indeed, he did walk up the ridge, but by then we were on the other side of a broad pond, approaching the safety of camp. We decided to move all the barrels and anything that might smell of food or us downwind a few hundred yards. The other option was to move camp to the other side of the lake, which left to our own devices we would have done. Rob held the fort until complete darkness set in at 11 p.m. A cold rain and penetrating wind lasted all night.

Loud, rude calls from sandhill cranes, literally inches from our heads, awakened us at 4:30 in the gray of dawn. Maybe our unwashed bodies alarmed them. We went back to sleep. The bear returned for another try at our camp in the morning, just after we left in the canoes. Strong headwinds gave way to absolute calm. It actually seemed hot at about 70°F (21°C). After fifteen miles, we were exhausted despite the ideal conditions. Rob was giddy, which he attributed to a kind of sensory deprivation from paddling big flat water. An island with three wonderful inuksuit was the destination for the evening camp, but it seemed never to get closer, engendering a strange otherworldly feeling. Joyce decorated Rand's hat with sticks so he resembled caribou, and with an increasingly bushy gray beard he was looking the part.

The island was a small rocky mound with few flat spots to place the tents. The inuksuit were outstanding. One was a bulky Baker Lake sculpture of a mother and child, while another was a masterwork sculpture of what also appeared to be a mother and child standing eight feet tall. The larger mother and child stood on the highest point on the island, with a grand view of the lake. It was just a grouping of rough rocks, perched and balanced one on another, but it resembled an excellent carving. The stones were different from each other and must have been carried here over winter ice, probably with dogs pulling a qamutiik. What an effort was required to make these rude stone sculptures, in the midst of nowhere now, but once in the center of their everyday lives. Our tent site was on a downhill slope next to an old

FIGURE 28 The most lovely inuksuk of them all, on 30 Mile Lake.

stone ring, leaving us at risk of sliding to the bottom of the tent. Sunset and sunrise were intense.

One of the fringe benefits of being sixty-nine and having to get up at night was that Fred got to see the tall mother and child at the top of the island in the light of early dawn. It was silent, except for an unexpected call from a loon. The inuksuk was visible in silhouette against pale purple and gray clouds, revealing only essence, the edge of mystery. Perhaps whoever created this work had in mind the power of motherhood, or the Great Earth Mother. In the disorienting sameness of white winter snow, she would be a beacon to travelers. Many inuksuit radiated power, but this one spread beauty across the wild Barrens. Our thoughts went to the Inuit who made this place home, embellishing it with art before they departed for good.

A couple paddled up in their spray-skirt-adorned canoe on another very calm day. They had come five hundred miles over about fifty days from the headwaters on Kasba Lake, and were cheerful and talkative. Our reaction was disappointment, because we felt entirely alone until then, and our peace was violated. They departed none too soon. Loons, ducks, geese, and jaegers enlivened our paddle. A tall Cleopatra's Needle–like inuksuk stood on shore close to the water, made of a single stone. Oral history has it that this was Ipjurjuaq's stone, which he placed here by himself to prove he was a stronger and better man than a rival, for the affections of his wife.[14] Some say the load was so heavy that his footprints remained on the tundra for one hundred years. Whether his footsteps are still there we did not know, as we paddled by without inspecting the tundra. The suitor apparently was impressed and gave up his unwelcome courtship.

Huge, wild Itimniq rapids lie at the end of Thirty Mile Lake, and we camped there to scout the waters and make plans. We took at least an hour and a half before dinner to scout the roughly six-hundred-yard-long rapid, which was adorned with shelves, islands, holes, and six- to eight-foot standing waves. Rob thought he saw a route through. Rob and Rand decided they would run the whole thing, but we were not certain that we could do the necessary forward ferry across the mid-portion to get to the only chute that seemed at all safe. Strength sometimes was a real necessity, and we undoubtedly were less strong than Rand and Rob. We had four children and ten grandchildren and we thought we would like to see them grow up, and also would enjoy the chance to grow old together. Odds of a spill were perhaps 20 percent, with the possibility of death from hypothermia and cold-induced bronchospasm if we spilled. With the path decided for the next day, Fred went fishing and caught a nice, fat Arctic grayling in the fast water on a dry fly, a parachute Adams, with a single cast to a rising fish. It was a tasty addition to a good dinner. We listened anxiously to the sound of the rapids as we tried to sleep.

After scouting the rapids again, we put on spray skirts, helmets, and life jackets and went for it. Both canoes did the upper part easily and as planned, a good thing since a miss could have thrown us into the big rocks and rock islands guarding the entrance to the lower half of the rapids. Rob and Rand helped us with a portage around the lower rapids. Perhaps we should have put out into the water below the rapids in preparation to pull them and their gear out of the water, but we elected to stay on shore and take pictures as they went through the maelstrom. They almost disappeared behind standing waves, and alternately pitched up and down as they coursed through very big water.

Rand recalled his feelings about running these rapids:

> Here we were on the Kazan, far north, in a land of dreams including the one about doing big rapids. Rob and I knew we paddled well together with strength. We had to paddle across the river and enter an eddy formed behind a large rock

FIGURE 29 Rand and Rob in really big water on the Kazan.

structure. The key was getting into that eddy smoothly and following it all the way up to the top of the rock, exiting without being flipped, and darting across the main stream to get lined up in the smooth flat fast current to the left of the huge standing waves. Doing it meant putting on all available gear to survive a dunking in the icy cold river. Rob put on his wetsuit, but I had only lots of cold-weather gear. Whether this was a good plan faded instantly as the canoe launched into the roaring embrace. Either it was going to be a fairly smooth and fast run, or we would be into the water through the rocks without the canoe. We concentrated completely, and it all worked perfectly. We experienced total exhilaration.

We saw increasingly large groups of caribou, meandering along the shore in groups of perhaps one hundred each, composed either of bulls or cows with calves. We probably could have walked among them, so peaceful were they in browsing and napping on the tundra. We made camp near Piqqiq, another famous crossing site with many old Inuit artifacts, above yet another but much easier set of rapids. This site, as every other site from Thirty Mile Lake onward, was on the south bank where the great majority of the old Inuit camps were located—a consequence of the traditional paths taken by the caribou during their crossing of the river. A short walk resulted in discovery of many old tent sites, some with a double ring; we wondered about the purpose of the ring within a ring. Did these designate storage areas, or play areas for children? At the top of a hill overlooking the river was a great stone-walled enclosure (Utaqqivvigjuaq [big waiting place]), once a hunter's blind. Legend has it that this structure was built to be a sort of prison, in which to trap and kill a particular man whose behaviors had upset the community. He allegedly crawled over the top, however, and escaped in his kayak.[15]

About 8 p.m., after supper while we were sitting in the bug tent, Rob saw caribou on the ridge behind us, moving in a steady stream over the crest. The relatively rapid and seemingly purposeful movement of the caribou caravan included bulls, cows, and calves. We grabbed cameras and binoculars and hiked across the tundra and up a hill to get a better vantage, aiming to see the spot from which they were coming. What we saw was amazing: a long line of migrating caribou, extending many miles, as far as we could see, and we could see a long way. They just kept coming, with no beginning and no ending, for the two and a half hours that we watched. Other than a few whispered words, we were utterly silent. The caribou were in a thin line, but very closely packed for the most part, with no loitering or grazing. A few big old bulls straggled, laboring to keep pace. In the soft, warm evening light, with a low sun that remained just above the horizon for hours, as is the mode in the Arctic, the entire line of caribou was colored gold. We could hear

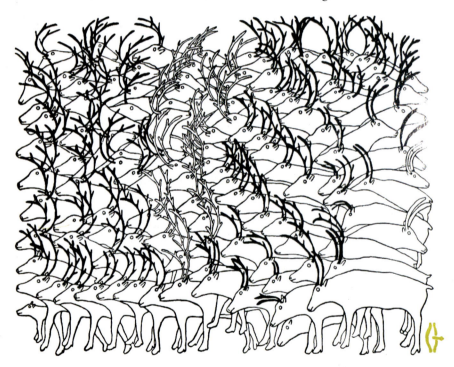

FIGURE 30 Ruth Qaulluaryuk, *Hundreds and Hundreds, Herds of Caribou*, stone-cut print, 24/40, 1975, Baker Lake. The print takes us to the bluff above Piqqiq on the Kazan, where we gazed for hours at the herd in migration.

them fording a creek. Rob sneaked closer and counted them, passing him at a rate of one a second. We heard the click and clack of their hooves in the distance. As they crossed over a ridge, their small figures were etched against the sky, and the hills seemed to be moving. Who knows how many there were? We estimated we saw at least ten thousand. We saw neither the beginning nor the ending. By morning it was over. There could easily have been several times ten thousand caribou on the march. They were on their southern migration, returning with their calves to the tree line for winter protection and food.

The great migrations of the caribou herds are not seen often. Indeed we encountered no one who had actually seen a migration, other than Alex Hall years ago on the Thelon. He had seen the big migrations many times, in herds of up to two hundred thousand, and had stood among them as they passed, grazing and nursing, ignoring his presence, taking over twenty-four hours from beginning to end. What we saw paled somewhat in comparison, but it was nevertheless stupendous. These migrations are smaller but similar in scope to the wildebeest and zebra migrations on the Serengeti. Unlike Africa, on the Barrens there are no drivers in four-wheel-drive trucks guided by radios to get one to the animals, only self-powered canoes and luck. Rob said it was one of the greatest things he had seen in his life. We agreed. Here was a primordial phenomenon, unchanged for millennia and yet threatened as big companies prepared to develop uranium, gold, and diamond mines, and to drill for oil. We had the very good fortune to encounter the migration in the golden light of the end of day. We hoped to see it, in our innocence, and we did see it. Sweet sleep . . .

> *Glorious it is to see*
> *Early summer's short haired caribou*
> *Beginning to wander.*
> *Glorious to see them trot*
> *To and fro*

> *Across the promontories.*
> *Seeking a crossing place.*
> *Yai-ya-yiya*
> *Glorious it is*
> *To see the long-haired white caribou*
> *Returning to the forests.*
> *Fearfully they watch*
> *For the little people . . .*
> *Glorious it is*
> *When wandering time is come*
> *Yayai-ya-yiya.*
> INUIT SONG[16]

Only a short paddle of four miles brought us to the rapids running into Kazan Falls (Qurluqtuq). A roar and rising mist declared that these rapids were different. The rapids ran directly into the falls, but if we did not run part of the rapids above the falls, the portage would be that much longer. We ran the top half-mile before pulling into shore with a nice tight eddy turn, and unloaded. The portage was about a mile and a third. Rob did some of the heaviest work, taking the two canoes. It was not easy to discern a trail across the tundra and rocks. A wrong turn took us to a side gorge that was difficult to cross. Working in stages, we got everything to the new camp below the falls in three round trips. The barrels were only thirty to forty pounds now, as much food was gone. The tent bag, however, was still a beast. Joyce declared privately to Fred that she was amazed to see him "prancing" with the load. That sounded like hyperbole, but Fred did feel great and was not particularly tired, and grumbled less than in the canoe when winds made it difficult to keep the canoe on track. Joyce carried her full share, a blue barrel of food or personal gear, plus assorted loose items. We made camp on a flat area near the river below the falls. Amazingly this was our last camp, although we were still thirty or forty miles from our planned takeout at the mouth of the Kazan where it entered Baker Lake. The spirits of the skies paid us back for our easy crossings of Forde and Thirty Mile lakes by sending blistering winds which started that night.

The wind was up on the morning after the portage around the falls, and whitecaps were all over the bay. The gale was especially bad on our second full day at this site. The bug tent blew down first, and was moved to a protected spot behind a small rocky promontory. In mid-afternoon Rand's tent blew up and away with all his gear inside, despite careful staking and placement of rocks on all the stakes. It tumbled into a little pond, which probably saved it from going forever across the rocky tundra. The tent was cut up, but it was repaired with duct tape. Rand remained calm, a real stoic—behavior much to be admired out here. He had the foresight to wrap everything in plastic bags in the tent, and his gear was not too wet. His tent and then Rob's were moved next to the bug tent, in a spot deeply rutted by caribou tracks. Rob wanted to move our tent, too, but after looking around and procrastinating, we stayed put. Our tent was well staked, held by big rocks at every stake, and was very tight. The tents were Eureka K2 XT that had been used on Everest and could take strong winds—we hoped. We were on a fine flat site, and had no rocks under our backs, whereas the other site was certifiably awful. The tent bent and shook in the raging wind, but it held. We got up repeatedly to tighten lines and check stakes.

Our breath was frosty, and we needed five to six layers, wool caps, and gloves to stay warm. Joyce, Rob, and Rand had winter parkas or down vests that they donned on top of several other layers of clothing. It looked like it could snow anytime. Snow in August is not at all unusual there. We walked around, fished, read, ate, and took naps. It was warmer in our bags. Our reading materials were long ago exhausted, but there was comfort in being there together, tucked inside the tent, waiting. We both remember it as a surprisingly peaceful and somehow rewarding time. We were rank and damp, however, and could smell each other downwind at sixty feet or more. We walked. The dark walls of the gorge were colored a bright orange in places.

Flowers were colorful on the steep slopes in the gorge. The tundra was about to ripen with all sorts of berries.

The loud flapping noises of the wind on our tent drowned the roar of the river, which exited the narrow sandstone gorge with huge volumes of water just below our tents. There was good fishing in eddies below the fastest water. Our eight-weight fly rod could not handle big fish in this water and wind. We caught beautiful lake trout from six to about twenty pounds with either fourteen- or twenty-five-pound test line on casting rods. Most of the fish were released to swim again, but several made it into the frying pan. By the end we needed trout for food, as we got down to just starches before we finally left.

A stone cairn above the falls held a small book with entries from previous paddlers. In the light drizzle of a dreary day, we found we were the seventh group down this year. Most trips did the whole 530 miles from Kasba Lake, and none used guides that we could see. We did not want to ruin this precious little book, and did not attempt to read all the entries. A year later we found a book that contained a photocopy of every note in this cairn, and also the cairn at the Hanbury-Thelon junction.[3] There were many entries by Alex from the Hanbury cairn, but none from people we knew except by reputation from the Kazan cairn. Some people saw a lot of animals, comparable to or rarely better than what we saw, but most saw far less. Some had seen ermine, snowy owls, and even wolverine, none of which we encountered. We wrote a brief note to accompany this distinguished list. Those notes were all earned by sweat except for one left by someone who dropped in by helicopter. Shame!

By the third day the storm rotated and the skies cleared, but the wind continued to rip at us. Caribou walked by the edge of the river and looked right in as we were having supper in the bug tent. Other caribou appeared on a ridge above camp. Recently dead caribou carcasses eaten out from behind testified that wolves were near.

Black and shiny musk-ox turds were everywhere, particularly on our camp site, so we knew they were close. We saw a herd of about fifteen musk-oxen in the distance above the falls, and another similarly sized group on the other side of the river well downstream of camp. A group of musk-oxen approached from the river below camp, walking slowly, seven animals in all. We stayed quiet and hidden behind rocks. They sauntered slowly up to and then just above our camp, a grand sight, with fine, deep-brown coats blowing in the wind, a bull or two, four or five cows, and a calf. Rob finally stood up and waved and made noise to keep them from suddenly walking right into camp, at which point they trotted slowly off. One of the notes in the cairn told of being charged by a bull.

An Inuk guide was scheduled to pick us up with his twenty-foot boat at the mouth of the river, and we thought he might be able to drive all the way up to the falls. Rob called him by satellite phone. He tried to get across Baker Lake on the fourth day of the storm, but was forced to go back. Moods deteriorated. We ran out of tea and coffee. Rob's previous record for consecutive wind-bound days was three, but we made it to five. Our flights from Baker Lake left in two days. Rob called Tim to ask whether he could pick us up, but got a noncommittal answer. We found a good, small landing strip close by with hard, flat ground and few stones. Another day passed.

At the end, a group of men from British Columbia caught up to us. They had been traveling for over fifty days, encountering terrible ice and black flies, but saw few caribou, missing the migration entirely. They were late and likely to miss their flight from Baker Lake, and opted to try to get on the bush flight out with us. The last morning on the river, our nineteenth day, was bleak, cold, and overcast with a low ceiling. The weather started to clear, however, and soon we heard and then saw the little Otter flying low. Tim swooped in to inspect the site, passing amazingly close—we felt like ducking—and then landed. Rob got in last and from the rear kept exclaiming how much we all smelled, and Tim agreed, opening his window. Our load was heavy, but we pulled up over the edge of the river in time, and flew back uneventfully. Musk-oxen and caribou were seen again from the air.

It was good to get back, although we all had trouble sleeping in a bed once we got back from eighteen nights sleeping on the tundra, where we slept like babies. We were happy to remove the sulfurous socks and to take a shower. Rand was really glad to see Adrienne, and she him. She had flown in three days earlier to greet us on our planned return, and by now had thoroughly visited the stores and sites in town while waiting for us, and had spent a day at an Inuit hunting camp outside town arranged by Rob via satellite phone. At meals at Baker Lake Lodge we met another group that had returned from the Back River, and learned three of their people had been burned, two badly, when a camp gas cylinder exploded. We also heard of an evacuation of a group of seven from the Back River due to hypothermia after a spill. There were stories of a young man who was mauled when he happened to meet a grizzly coming up the opposite side of a ridge near the Kazan; he was evacuated to a hospital, but survived. His same group also lost a canoe permanently when it blew away one evening in the gales that trapped us below Kazan Falls. Our trip seemed almost tame.

We had a big hug with Rob at the airport. Later he wrote in his newsletter, "I was amazed at the scenery but more so the joy at which the members of our happy group took it all in, becoming a part of the landscape that few can comprehend and fewer love as much as they do."[17]

Although we paddled only 130 miles, we traveled through places where the impact of the Inuit was still fresh. The great sculptor George Tataniq lived on Forde Lake, where we saw the bull caribou swimming the lake.[12] The printmaker and wall hanging artist Irene Avaalaaqiaq grew up in a camp near Piqqiq rapids.[5] Her grandmother may have told her stories about myths and traditions, and she may have fantasized other children as playmates, right where we saw the caribou migration.

Of all the remarkable memories, the caribou migration was most powerful. Perhaps our reaction was no different from the cave painters in prehistoric France and Spain, who etched and painted images not of humans, but of bison and bulls, horses and cave bears. They depended absolutely on the great animals, and clearly had reverence for them.[18] Our genes are directly descended from these early modern humans, and we seem to be hardwired with an innate sense of awe for the great migrations, and of the great open spaces on which the migrations occur.[19] Certainly the Inuit still depend in many ways on the idea as well as the substance of the caribou.

Under big skies and on big water on the immense open tundra, we felt ourselves to be small. In the cold and wind, we felt our lack of power to control things. In the pastel skies and transient but beautiful wildflowers, we saw beauty. In the migration, we saw the continuity of life. It was a spiritual experience, in the cathedral of the Barren Grounds.

FIGURE 31 Caribou are curious. Two young bulls and a cow walked right into camp along the Kazan.

8

WILD RIVER: TO THE BACK RIVER

Cold and mosquitoes,
Mosquitoes and cold.
Thankful are we
That these two plagues
Come never together!
INUIT SONG[1]

JULY 2006 It sounded like rain, but it was only the steady pinging of the black flies against the tent, a steady beat. The dome of the tent showed a soft glow from the late evening light, and we were happy to be inside. The day before, Fred had a coughing fit from inhaled mosquitoes. The usual trick of trapping the mosquitoes in the front of the mouth and then spitting them out had not worked. We were halfway into the long-anticipated paddle with Alex Hall on a tributary of the mighty Back River, which we will call the "wild river," in deference to Alex's wish for anonymity of his routes. The bugs were letting us know who was in charge. Tundra fever still had us in its grip.

The wild river is not far from where the Utkuhiksalingmiut Inuit once lived, to the east down the Back River closer to Chantrey Inlet. Alex described

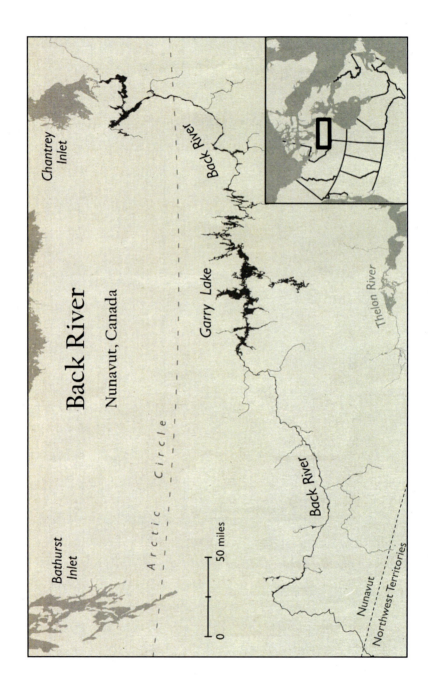

this as the best of all his trips but potentially the most demanding. The wild river is subjected to more northern storms than any of the other rivers that he paddles, and is notable for rapids throughout its 130 miles.

We met most of our colleagues in Edmonton on the way to Fort Smith. Brother Rand was aboard again, our stoic survivor of the Kazan gales. Joyce was the lone woman among ten canoeists, almost all of whom were strong and fit middle-aged males. We wondered what they thought about us, a pair of not very big seventy-year-old elders. One commented that we looked fit, although it sounded as if he were trying to reassure himself that we were not going to be a problem.

Ice collected on the plane during the flight north from Edmonton. The pilot told us that this was just the usual little icing that happens in the North, not to worry. Not surprisingly, the pilot was right. Alex met us in Fort Smith and escorted us to the Pelican Rapids Inn. A day in town helped us get acquainted with each other. Sadly, our Metis bar hangout had burned down, and the superb little bookstore with a fabulous collection of books of the North had changed hands and had dispensed with the books we loved. Change is inevitable, but we wish the world would slow down. Everything essential was repacked in double-plastic-lined canvas Duluth packs. Our gear included wading boots with full felt soles plus knee-high three-millimeter neoprene socks with fleece lining, in preparation for hours spent wading over slippery boulders in cold water.

Two years ago the spring was very late and everything was closed in ice, and now it was the exact opposite. The ice went out of the Slave River two weeks earlier than the previous record. We expected more flies and lower water than usual for this time of year. Rand rode in the old (circa 1948) Beaver with all the gear that it could carry, hopefully well prepared for a four-hour trip with no pit stops. The rest of us flew in a Cessna Caravan, a bigger and newer single-engine plane. We covered over 350 miles, northeast from Fort Smith to the wild river, in about three hours, passing Rand in the Beaver. On the way we saw not a bit of ice, unlike two years before when the big lakes in the Barren Grounds were ice-bound. Crammed into the plane with a lot of food and gear but not knowing each other well, we were close yet still distant from each other. The sun sparkled like thousands of diamonds on the lakes below. We landed on a wide spot in the river, near its origin, surrounded by tundra and low hills.

Alex read us his rules again in their entirety. To our chagrin, during the next eleven days, we violated two rules: carefully assemble the poles before erecting the tent, lest the aluminum poles snap, and do not get sand on the bottom of the packs. Departure was put off until the next day. Sik siks came out of their holes and cavorted in the sun, aware of us but looking happy. Blue lupine and the much less dramatic liquorice-root plants were in bloom on a hill above camp, and the lupine were scattered about in great patches. The grizzlies as well as the Inuit loved to dig liquorice-root plants, the potatoes of the North, to eat their starchy tubers. We caught fish and made camp. Evening light was soft and a bit warm, in this peaceful, totally secluded, and wild corner of the planet.

In the morning we loaded up and checked the balance in our canoes. This trip would have a full complement of five canoes, suggesting less sense of isolation compared to our earlier trips, in which we had the good luck to have only two or three canoes. Joyce paddled bow, as she had previously, leading to Fred's nickname of "Stern man," which was not necessarily a compliment. If Joyce seemed to have a more boring job in flat water, she was the key person on this trip. In fast water the stern man had to trust the bow person's decisions, since she was in position to see rocks first and was responsible for initiating urgent steering maneuvers to avoid hitting rocks. When she did a draw, her stern man needed to do a pry without questioning the decision, to keep the boat moving vertically down river. We did not want to broadside into a rock, where we could be trapped and overturned by powerful water coming down on us. Capsizing the boat would not be a good thing even though everything was carefully strapped into the canoe. Alex did not use spray covers on his trips, worrying that one could be trapped under the cover and drown as a consequence.

The water was markedly lower than on Alex's previous trips on this river. Much of the first day on the water was spent alternately paddling through or dragging the canoes over narrow rock gardens. We pushed, lifted, hauled, and generally scraped the canoes terribly. Thank God for Royalex canoes with fiberglass strengthening; the Pak canoes of last year would have been shredded. An unplanned short portage was necessitated by an impassable ledge. Once we got into deeper water the paddling was a sweet pleasure. Some rapids were quite technical and challenging, requiring coordinated and fast teamwork in bow and stern (*Draw! Pry!*). We were not always stellar, but no worse than the others. One of our most experienced team members taught us the theory and practice of a good cross draw: keep the arm holding the top knob of the paddle down, and use it as a forward-pushing lever to drive the paddle. We were glad to be students, and thankful for the occasional tips on lining techniques as well. As intermediate-level whitewater paddlers, we listened carefully to the experts among us.

Rose-colored dwarf Arctic fireweed emerged as we moved downriver, occupying great swaths of the sandy soils by the side of the river and the gravel- and boulder-strewn fields. The blue of the lupine and rose of the fireweed were visible at a distance from a moving canoe, but one needed to get down on hands and knees to appreciate tiny white bells on Arctic heather, and equally tiny pink blooms on low mats of moss campion. Patches of mountain avens raised their white flowers to the sun. Yellow cinquefoils dotted the landscape, and clumps of white prickly saxifrage were common. Willow catkins were in bloom, and white Arctic cotton swayed in the breeze in moist places. Large-flowered wintergreen was seen occasionally. Bearberries and cloudberries, Labrador tea and other flowering plants added to the richness of the tundra. This was the peak of Arctic summer. We saw only one small, lonely clump of stunted spruce during the trip.

Sandy hills and eskers were replaced by brown hills, and then by black rocks and low sandstone cliffs as we descended the river. There were occasional patches of snow on north-facing slopes. Our moods changed quickly

FIGURE 32 Arctic lupine by the wild river.

in concert with the changing sky, first blue then cloudy gray, but the weather was calm, and we had no wind delays. At times the water was wide and slow, and we could bask in the warmth of the sun. There is no pleasure greater than lying back, stretched out fully, relaxed in the extreme after paddling hard.

Most of the trip was down fast water. We all were in up to our hips periodically getting out of messes, or through narrow spots rife with boulders. Once we saw a round rock too late and hit it mid-canoe; by luck we rolled off rather than over. As the river grew bigger, the rapids became deeper, requiring faster runs through waves and fields of boulders that sometimes seemed to emerge out of nowhere.

"Wow, didn't see that one…or that one either" (Fred).

"But I saw them and you were fine" (Joyce).

We passed the boulders very quickly, paddling hard to keep traction in the moving water, leaving time to think later. It was exciting.

Alex is all about safety, and we were entirely confident in him. Frequently we scouted the river, walking the banks down to the rapids. The path usually was near shore's edge, where the water was shallower and the waves smaller.

Sometimes the desired chute was extremely narrow and had to be hit precisely. Seeing the path as we plunged downriver could be difficult; landmarks in the river were harder to identify while paddling through foam and turbulence. If the waters looked unsafe, we lined down instead of paddling. We must have scouted fifty rapids throughout the trip, running about half of these and lining the others. The remaining one hundred rapids were run without scouting them. No one spilled, and no canoe was damaged.

One of the rapids had powerful standing waves on river left, and an impassable rock garden on the right, with no slot between. The shores were a massive field of big boulders. Portaging would have been nasty, and lining was impossible. We walked the canoes through, in knee- to hip-high fast water. It seemed to take forever, as those ahead of us negotiated the gauntlet. One of the canoes got turned around somehow, aiming backward. Footing was perilous, and the water surged around us. The lead canoe got into trouble, and we all had to wait for what seemed many minutes as they tried to figure out how to solve a quite hazardous predicament. It seemed we might have to jump into our fully loaded canoes from thigh-deep fast water to exit the rock garden. Doing that bit of acrobatics at age seventy was a trick we wished to practice before doing it for real. We found another way to squiggle through, with the help of colleagues who waded into the torrent to help. The group had its finest moments here, keeping calm and working together.

Our confidence was greatest after a day of running many fast rapids without hitting a single rock. The next morning brought a fast end to complacency when after paddling only five minutes, we ran flat aground in the middle of the river on a large table rock. We saw it, but reacted poorly, being lazy, misjudging the force of the river, and miscalculating the vector of our path. Momentary panic ensued. Fred got a foot out, found a solid foothold, lifted and pushed, jumped back in, and got us going with a momentary wobble but no water in the boat. Others saw the gaff and commented at dinner. There were no secrets on the river.

Joyce was flattered by a comment made to Rand by one of the young, strong men that he was able to keep paddling when fatigued because he saw Joyce was still paddling hard. Others commented that we looked good together, in nearly perfect rhythm, paddles feathered into the wind and just above the water, stroking smoothly with no fuss. Rand was saddled with a canoe partner who insisted on paddling stern and on the right side 95 percent of the time, even though the bow man was switching sides as was both usual and necessary. Seeing both of them paddling the same side was a sight, particularly when the stern man continued to adjust the path of the canoe with braking strokes, causing Rand to use tiring power strokes to regain momentum. Always one to suffer quietly, Rand merely persevered, making no comments.

At its best, lining has grace and rhythm, with the stern man and bow man playing complementary roles. The stern man held the stern line taught until the bow person was well ahead of the bow. The bow had to remain farther out in the river than the stern, or else the water would try to sweep the stern out and around the bow, putting the canoe at risk of filing up and capsizing. It was important to keep pace with the canoe and to step promptly over slippery boulders, eyes on the canoe rather than feet. Not watching anything but the canoe in the river could be hazardous. Once rocks well out in the river required the stern man to push the canoe out far as possible, letting go of the stern line and trusting the bow man to hang on and bring the pirouetting canoe back in safely. It was lovely to see this unfold, in what seemed slow motion. Joyce was impressed that Fred trusted her to hold on to the bow line, because if she failed, we literally would have been "upstream without a paddle."

The river was full of boulders; the banks were lined with them, sometimes in fields a hundred yards wide or more. At times the campsites were reachable only by carrying all the packs and canoes up over the boulders. When the boulders were wet and slippery, and also unstable, walking over them could be dangerous, with or without a load. Everyone fell hard, often several times, resulting in deep bruises for many.

A small plane surprised us, flying low, most probably company people looking for places to mine and despoil, and most certainly perturbing our sense of isolation and quiet. Alex yelled up at them, saying he would shoot them if he could, repeating the thought a day later for emphasis. A figurative but not literal threat, we were certain, but powerful nonetheless. The plane followed a long, low, sinuous path along the river, moving sideways like a snake gliding across the desert, evil in its intent. We understood Alex's feelings.

One thing about Alex: He is a man of conviction, and has long waged a fight to save the wild lands of the Barren Grounds and the NWT for future generations. He wrote us later, relating how he and his allies were having increasing success in preserving the lands around the upper Thelon from mining. Alex cares about the wild places, and knows them like few others. We were privileged to paddle with him. He is an heroic figure in the fights to save the boreal and tundra wilderness for posterity, but the forces opposing him are strong, fueled by greed and the pressures of huge and growing human populations for natural resources. Every wild place is under pressure, and many or most are already seriously degraded. The Barren Grounds experience is so powerful because the Barrens are still largely intact, a huge primeval place that humans have not yet ruined.

Few caribou were seen, about twenty in total, although Alex encountered large numbers of caribou here in previous trips. We were spoiled the year before on the Kazan, where we saw them by the hundreds and thousands. The unpredictability of the caribou migrations added somehow to their mythic power. Perhaps the lack of caribou on the wild river was not unusual in historical terms, since we saw no evidence of former Inuit camps—no inuksuit, tent rings, or any similar artifacts. For certain, had we been farther down the Back River toward Garry Lake or Chantrey Inlet, we would have encountered many former Inuit camps.[2]

Two white wolves appeared on the hillside on the first day, and one dashed down and grabbed a flightless goose, carrying it rapidly over the top of the ridge, presumably headed for a den and pups. Three more were seen near the water later, looking shaggy in their partially shed winter coats, and behaving furtively. They must have been hunting for flightless geese also, an easy target and welcome in the absence of caribou. This was the moulting season, and geese were relatively helpless. We walked to see old, inactive wolf dens on a couple of occasions, with one of the group falling into an old, unoccupied wolf den, which could have but did not break his leg.

An otter swam across a pond, which caused considerable discussion because Alex never had seen otters this far north. That otter was not a happy sight, more evidence of global warming. Otter north of their traditional boundaries, and surveyors in low-flying planes scouting for new mining sites, were bad omens for the future. A pair of Arctic foxes appeared above camp, one cinnamon and one black. Birds of many kinds were seen, including common and yellow-billed loons, short-tailed jaegers, sparrows, larkspurs, seagulls, geese, and a single golden eagle. Eggs were discovered laid on rocks or under willow branches, but we knew too little to identify them.

After a long, arduous walk across bogs and hills and through willow thicket, following a musk-ox we spotted from the canoes, we finally caught up, surprising the massive old bull who had laid down to rest in a warm, sandy spot in a hollow behind a hill. There were too many of us for him to do anything but retreat to the top of an esker, where his long coat streamed in the wind. It was thrilling for those who had never seen a musk-ox, and satisfying for those who had. This great specimen might be the last one we would ever see, for all we knew.

> *I wish to see the musk-ox again.*
> *It is not enough for me*
> *To sing of the dear beasts.*
> *Sitting here in the igloo.*
> INUIT SONG[3]

Spotting a sow grizzly and two small cubs coming across the boulders and through huge patches of fireweed downriver, we got out of the canoes on the shore opposite the bears and watched. They moved very slowly. She entered the river, but her little cubs seemed terribly nervous and did not follow, so she had to return and admonish them. Curiously, she had a very dark body and almost white head, as did one of the cubs. The second time they all swam across to our side. They headed upriver toward us, and we got into the canoes, pushing off a bit from shore, hoping to see them, while avoiding a confrontation. Very soon the two cubs showed up on the boulders, curious and looking like toy bears in a shop. They sat down and were totally unafraid, hugging each other while staring at us. Mama Bear lost sight of them and became alarmed.

Rand recalled this moment: "She was close. She looked directly at us and gave a huge roar. This was a roar that made it indelibly clear who was in charge. It was the kind of roar that defines the wild in wildness. Not one of us moved. We knew exactly where the cubs were; we were looking right at them." Soon she saw her cubs, gathered the little miscreants, and herded them over the ridge.

Our only other bear encounter was via a story Alex shared at dinner, a scary confrontation he and his son had a few years prior, when a grizzly bear started to maul their tent but left when Alex shouted at it from his sleeping bag. That story did not make it into his book.

On the day before we were to be picked up, we made camp at a sandy site, hauling our gear a long way to a plateau well above the river. Above us towered an esker that was perhaps 350 feet high. There was a lot of caribou hair on the water's edge for the first time, and nearby there were caribou tracks only a few weeks old, testifying that a small herd had been around recently. We hiked up the huge soft esker, aiming to find a wolf den several miles distant. A big blow came up suddenly, catching us in the open, but subsided as quickly as it started. Fred lagged behind. His right knee was stiff and swollen and his lungs were burning, his breathing heavy and labored. He was always pushing the envelope when it came to functional lung capacity, and uphill climbs were difficult. Joyce suggested it might be best to admit the obvious, and let the others go without us. We picked our own route back, and wandered along the top of the esker. The sense of being alone was special, spiced a bit by stories of past bear encounters on this same spot. Joyce clutched her bear whistle ready to blow at sight of a bear. Flowers peeked up here and there in patches. A grand panoramic view of the open tundra was deeply satisfying somehow. The tents looked tiny above the winding river, which gleamed brightly in the low light.

The early morning skies were wonderful, with low white cumulus clouds lit brightly by the very low sun behind them. Camp was broken, and we paddled down to a good takeout spot, with a broad sandy beach. A water surveyor's hut nearby was covered in sharp nails and spikes to keep marauding bears away. Wolf prints were outlined in the sand. The gigantic esker we walked the day before was outlined against a sky full of soft clouds, across the river. Before long, too soon, three planes arrived relatively close together, a Cessna 185, the Beaver, and the Cessna Caravan. The wind was up, and we had to tie the planes down to keep them from piling up on each other. Canoes disassembled, gear loaded, we flew out to deposit the canoes in the Thelon system for Alex's next trip. Scattered small trees reappeared, and we landed on a beautiful small lake lined by sand dunes and spruce trees, where fuel drums for refueling were hidden. The sun came out, and it was bright and calm. Beautiful lakes and rivers were spread out below us, literally hundreds of miles of them, sparkling blue amid the greens and browns of the tundra and yellow-gold of the winding eskers. Our minds were occupied with memories of moments of repose in the warm sun, the excitement of paddling through so much whitewater. There was no room for worries about the decline of the caribou. That would come later when we learned more.

To be exhausted by one's own labors, totally free of the cares of the routine life in the modern, complicated, overdeveloped Western world, is an exceptional experience. The loneliness and vastness of the Barren Grounds have an appeal that has to be experienced to be fully appreciated. The wind, cold, fatigue, bruises, bugs, dirt, and occasional danger are necessary parts of the whole. It was a long way back to Fort Smith, and a much longer conceptual journey back to the everyday reality of work and world events, to a world dominated by television, computers, cell phones, automobiles, and great masses of people. One of our goals was accomplished: to paddle three of the great rivers of the Barren Grounds, rivers and lands that helped define the lives of our Inuit friends. We were a team in the canoe as well as in the galleries. Sharing these memories with Rand was wonderful for brothers who had been separated by distance for most of their lives. If bad lungs and bad knees were to limit future trips, we squeezed in some marvelous experiences, just in time.

9

FORTY BELOW: IGLOOLIK

Birthed from laboring clouds above
Snow lies silently on the land
Like bedsheets of white,
Smoothed and tucked by unseen hands.

JEAN MCLENNAN[1]

JANUARY 2006 Our dream was to experience the snow, ice, and extreme cold of the Arctic. Because Fred might have trouble with his asthma in frigid weather, it was my [Joyce] duty to enjoy the cold for both of us. An opportunity presented itself in January 2006, when Inuit art-dealer-turned-tour-director, Carol Heppenstall, offered a trip to Igloolik—but for ten women only. The purpose of the trip was to see the sun emerge for the first time in over six weeks, the winter solstice; cause for a yearly celebration of renewal. Purchase of a calf-length down-filled parka with fake fur hood, and boots and gloves that were advertised as foot and finger warm to $-40°F$—added to my ski goggles, pants, face mask, and long underwear—gave me confidence in confronting anticipated low temperatures.

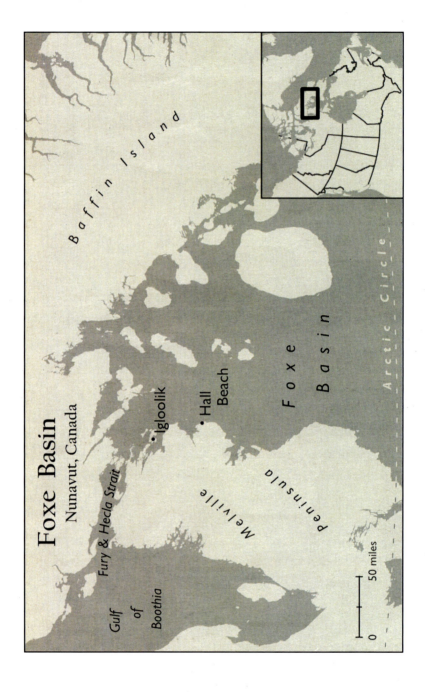

Igloolik was familiar to us through the film *Atanarjuat: Fast Runner*, winner of the 2001 Cannes Film Festival and Genie Awards. The vivid, long, dramatic chase of a naked Inuk over the frozen sea in that film was unforgettable. Pakak Innukshuk, a costar of *Atanarjuat*, was a resident of Igloolik. I anticipated renewing our friendship, initiated during Fred's and my first trip north. Pakak is directly descended from Aua, an Igloolik shaman whose hunting prowess, habits, skills, and beliefs were extolled in Knud Rasmussen's account.[2] Eighty years ago, Rasmussen learned from Aua of the Inuit taboos and traditions, some of which inform the beliefs and practices of older Inuit today.

Awaiting my flight north, an article in a Canadian newspaper caught my attention: "The World According to Fred." Frederick was described as a "peace ruler," and use of the name "Fred" was considered a stress reliever. Though shocked by this information after being married for some forty years to Fred, this article relieved any trip tension. Upon entering Canada, my backpack was searched; several security men smiled as they pulled from the pack my "sipping whiskey" silver flask deeded to me by my ninety-four-year-old teetotaling aunt. They sniffed it, removed it to the men's room, and returned it empty amid many sniggers from passengers as well as security agents. How many more errors in judgment would I make without my stress-relieving spouse?

Snow greeted me in Ottawa, as well as news that my luggage carrying all my cold-weather gear was lost. The concierge at the stately old Fairmont Chateau Laurier Hotel, rather than encouraging me about the odds of recovering my luggage, excitedly informed me that my roommate was a famous Canadian actress. Unrelieved, but trusting, I headed for the elegant Inuit art collection at the National Art Gallery of Canada, followed by a visit to the Chateau Laurier bar for a sip of peaty Cragganmore single-malt whiskey. Whatever happened, this Arctic trip was bound to be memorable.

From Ottawa, we flew to Iqaluit, capital of Nunavut, and on to Igloolik the same day.

Appended to my ticket in Iqaluit was a green stamp, a firm statement that the airline would not be responsible for inability to land on this small island due to bad weather. Located above the Arctic Circle at 69° north, Igloolik has been occupied by Eskimoan peoples for about four thousand years. Seals, char, caribou, polar bears, and large and reliable herds of walrus abound, providing the basis for life in this frigid environment. Although the sea ice breaks up in July, extensive patches of pan ice remain all summer, which caused whalers to avoid these waters. The result was limited contamination by southern culture until the third and fourth decades of the twentieth century.

Parry, Lyon, and their shipmates were the first Qallunaat to visit Igloolik, remaining over the winter in 1822–1823 on their second journey to try and discover the Northwest Passage. Fury and Hecla Strait, a narrow passage to the high Arctic forty miles north of Igloolik, is named for their ships. Parry's journal provides a detailed record of the dress, customs, and beliefs of the local Inuit.[3] In 1918, white missionaries introduced Christianity to Igloolik, followed by Rasmussen and his colleagues' anthropological studies in 1922.[2] Further contact with white men occurred after the establishment of a Hudson Bay trading post in 1936. Recognizing its mineral potentials, the Canadian government became more interested in Igloolik in the 1950s and developed a school, nursing station, and RCMP post. The fundamental independence of the Inuit in this community, however, was expressed in their 1973 and 1977 refusals to allow English-language radio and TV in the community.[4] Eventually their attitudes toward modern technology changed, and from 1989 to 1995 Isuma Productions produced sixteen Inuit tradition films, followed by other documentaries and eventually by *Atanarjuat*. The Inuit, themselves, always adaptable, brought the white man to their door.

The haze lifted as we circled above Igloolik and gazed upon an enormous expanse of white wilderness. There was no way to tell where land ended and frozen seas began; all was covered in either snow or ice, seen by the light of the moon. We were greeted by smiling faces peering from furry parkas. The warmth of our welcome displaced any shivers we might have felt, but outside the airport, the cold was startling in its intensity. Our small band of intrepid women huddled together in an unheated van carrying us a short distance to our small, motel-type hotel. This recently built structure had a rare northern amenity: a bath for each room.

With my winter gear retrieved, including a neckpiece made of the soft underwool of the *umingmak* (musk-ox), I was prepared for the temperatures, but not for the biting invasiveness of the wind. Taking gloves off for snapping just one picture was painful, making it difficult to record this fantastic experience.

It was dark outside. We noticed few stars, although they may have been obscured by the light of the moon reflecting off the snow. Wearing every possible piece of warm clothing, we crunched our way over the dry, snow-packed roads toward the contemporary mushroom-shaped Arctic Research Center and on to the comfy home of John and Carolyn MacDonald. Carolyn, a special education teacher, is director of the Head Start program in town. We were to observe one of her preschool classes later and hear of the extensive curriculum she and her colleagues established for Igloolik and hopefully for all of Nunavut. John is manager of the Igloolik Research Centre for the Nunavut Research Institute. The MacDonald family has lived in the North for decades.

John's passion is communicating in Inuktitut with elders of the community about Inuit myths, oral history, and lore related to stars and sky, much of which is recorded in his book, *Arctic Skies*. Inuit once used stars, along with their intimate knowledge of the wind and the contours of the snow, to help guide their travel in winter.[5] The far northern skies are different from those which people living in lower latitudes are accustomed to seeing. In Igloolik some stars (such as the Milky Way) are never seen, for they are always below the horizon.[5] The light is different also. Light is affected by the low position of the sun and by temperature inversions, resulting in weird phenomena, such as occasional double images of the sun or the moon, magnified images, or upside-down images.[6] John was recording the old navigational and hunting skills that allowed Inuit to survive the long, cold, and dark winter, and teaching these skills to the youth in the community.

The MacDonald home contained a vast library of northern books on the history of the area. On the wall was a stone-cut Inuit print by Simon

Shamaiyuk (1915–2002) from Pangnirtung. This print of a hunter pulling a seal along the ice was particularly interesting, because several years previous, I had purchased the print stone from which this print was made. Usually after a limited edition of prints is completed, the stones carved for printing are destroyed. This one somehow escaped destruction.

After thanking the MacDonalds for their hospitality, we walked quickly to our rooms, hearing only the sounds of our large clumsy boots on the crisp, dry snow (*puviksukartuq*). Our European friend exclaimed, "The moon looks like a bloody street light." The wind was blowing, and the temperature was fast approaching $-52°F$ ($-47°C$). As I readied for bed, curiosity made me turn on the TV at around 11 p.m. A message to all parents stated, "Do you know where your children are right now?" I couldn't imagine children out this late in temperatures this cold.

Rachel Uyarsuk, the eldest person in the community, told us one myth of the origin of the moon and sun, supposed rivals in the sky. Rachel spoke in Inuktitut, the rhythm of which gave a more spiritual reality to the tale, which has many iterations. The following story is one of them.

A brother and sister lived together. Each wanted the other to marry, bringing another worker into the family unit, but neither wanted to do so. One night the brother forced his sister to submit to him; because it was dark, she could not identify the man and painted some soot from her cooking pot on his nose for future identification. When she realized her assailant was her brother, she informed him she was leaving forever. Holding a lamp to light the way, she ascended skyward, becoming warmer as she ascended, ultimately becoming the sun. Her brother was distraught and lonely, so he followed his sister, soon ascending to the heavens to become the moon (the man in the moon), and to this day still following his sister the sun.

The hamlet's shore was recognizable only by the bow of a boat peering from the frozen seas. The snow engulfed us, the wind pierced our coats—all so different from the whisper of winter and my rare encounters with snow in North Carolina. One of our group's Canadian poets described the setting:

FIGURE 33 The moon over the endless frozen sea ice as seen from the shore of Igloolik.

> *Wind breathes onto the scene.*
> *The snow, overwhelmed, succumbs to this force*
> *And becomes sculpted, encrusted—*
> *Though the brilliance of frozen diamonds*
> *Still is present—*
> *Their multi-carat perfection blinding the eye.*
> JEAN MCLENNAN[1]

    A day was spent at Isuma Productions, an impressive group of filmmakers led by the creative energy of Zacharias Kunu. Trying on the sealskin outfits of the participants in *Atanarjuat, the Fast Runner*[7] transformed us Qallunaat into formidable-looking Inuit. Igloolik women sewed the clothing for the film, and their pride in this accomplishment was obvious. The stitching, colors, and textures of the fur were exquisite. The Inuit enjoyed our enthusiasm, joined our laughter, and urged us to try on the winter caribou clothing worn by Pakak in the new Isuma film, *The Journals of Knud Rasmussen*.

Settling mining rights is a critical issue for the Inuit and was inextricably linked to the acceptance of Nunavut as a territory in 1999. We were fortunate to be introduced to the chief arbitrator for Igloolik and heard of his extensive and prolonged work on land rights. The people of the NWT had achieved voting rights in 1962, but it took thirty-eight more years for the Inuit to gain control of their own health services, education, and administration of justice. Hearing the saga directly from him, we began to further appreciate the intelligence and emotional energy as well as the restraint the process required. The present goal of the Inuit arbitrators is to achieve provincial status for Nunavut, which they suggest may take until 2020 to achieve. Another Inuk presented pictures of a few Inuit traditions and food preparations: for example, walrus meat placed in the walrus skin, laced together tightly with sinew, buried within a pocket of stones, and allowed to ferment. Graham Rowley described the resulting green, cheesy delicacy on his dogsled trip to Igloolik in 1936.[8] Fermented walrus undoubtedly is an acquired taste.

Meeting the elder Pauloosie (the one who wears caribou) Qulitalik, a founding member of Isuma Productions and cultural advisor to *The Journals of Knud Rasmussen*, offered a unique opportunity to ask questions of a wise and experienced Inuk. Queried about his perspective on the apparent changing weather and the melting of the ice, he replied, "We must not linger on negative factors because they will become larger problems if we do…. The weather is always changing." He asked, "What do you really want to know?"

One of the women answered wisely, "To know another culture is to refine one's understanding of one's own culture."

A few days later Pauloosie spoke at a meeting about the designation of narwhals and bowhead whales as endangered. Speaking to the government representative, he repeated his disdain for comments about changing weather and its effects, and of the southern Canadian's need to "control," sadly adding, "I feel useless, no one listens."

On our walks to the Co-op or Northern stores, we noted Inuit wrapped in reams of warm clothing, not the anticipated caribou or sealskin coats of some in Pangnirtung, but handmade, multicolored amautiit and southern jackets. We were guided by the pale moon and the hint of an emerging sun. A vast expanse of whiteness extended from town well into frozen Foxe Basin, and we could see for miles. Dots of darkness peppered the frozen landscape: a dog wandering, seeming lost but knowing better than we his path; a skidoo returning from seal hunting or from Hall Beach, about forty miles from Igloolik. The spirituality of the experience eluded our cameras.

Rosemary Clewes, a member of the trip group and well-known Canadian poet/writer, expressed our growing appreciation of the light: "When I'm away from the North, I dream of the light.... I am bewitched and thinking about spirits; light's disproportionate presence tantalizing my mind. Air's dust-free clarity makes far seem near. I can touch the top of the mountain or walk ten miles in a minute, while at the same time the diaphanous space stretches the meaning of infinity."[9]

The Inuit were quiet, dignified, and succinct. When asked how many char he caught that day, one Inuk replied, "Enough for the day!" I was reminded of Alvah Simon in *North to the Night*. With the winter freeze descending at latitude 73° north, he searched for a winter anchoring site for his and his wife's sailboat. Simon asked an Inuk for advice. The Inuk simply smiled. Simon reported a feeling that I came to understand: "Perhaps the Inuit will allow a few of us in the Arctic for our sheer entertainment value."[10]

Would you believe that a North Carolinian would be in demand as a teacher in the Igloolik Arctic College? It all started with a casual breakfast on a typically indolent morning. An itinerant teacher for the Arctic College learned I was from North Carolina, and asked if I would read a novel about the Civil War and discuss it with her students the next day. *Property* by Valerie Martin was set in 1828. The book focused on a slave owner and his extended family and emphasized the effect of power and authority on all his contacts. The universality of people and their goals was a clear agenda for the class: understanding the effects of deprivation and authority on individuals, the joy of meeting adversity and surviving, confronting the changing roles of women as workers, mates, and mothers.

One student, Jeena, stood out in her willingness to engage in open discussion. She declared that she always wore a mask; at home she ordered her children to get her a drink as the slave owner did, but she always obeyed her father and teachers just as the slaves responded to their master. Her admission supported the class agenda and led to a less-restrained discussion of the students' roles and interactions. After class Jeena—or "Mask from Igloolik," as she sometimes called herself—asked me to send historical novels and requested that I contact her by email. That request initiated what became a meaningful friendship, enabled by the miracles of computers.

Jeena had a steady partner and, at this initial meeting, three children. Her first child was adopted by her mother apparently because Jeena was in school. Adoption is common here, and may either be a practical necessity, a traditional courtesy, or a means of child protection. As a pediatric clinician, teacher, and mother, I was thrilled at their concern for all children. No Inuit child is unwanted, and none is allowed to be an orphan. When visiting Carolyn MacDonald's program, I was also moved by the appropriate care and instruction provided for a child with special physical and educational needs.

The sun was about to emerge for the first time in six weeks. Qamutiiks would carry us over the snow toward the horizon, where we would have a perfect platform to appreciate the storied spectacle. An Inuk appeared at our doorstep with a sled pulled by eight rambunctious huskies. It was dark and bitterly cold. Four of us clambered into a large box tied on the sled, wrapped ourselves in layers of caribou skins, and settled down. The dogs were run in a fan configuration, their breath steaming. From my reading I recalled that the dogs are given a break when traveling over rough spots of ice and snow by one person jumping out, running alongside the team. When the dogs slowed down over a rough spot, therefore, I jumped out, but fell ignominiously in the soft snow. By the time I righted my heavily clothed, cumbersome body, the sled had picked up speed. I could not keep up, impeded by my heavy boots sinking into the deep snow, my long coat preventing leg movement, and my goggles filling with snow. Attempts to catch up brought gales of laughter and provided our Inuit friends with the sheer entertainment that Qallunaat bring to the North. I remembered a song that Aua sang when heavy things need to be made light:

> *I will walk with leg muscles*
> *Which are strong*
> *As the sinews of the shins of the little caribou calf.*
> *I will walk with leg muscles*
> *Which are strong*
> *As the sinews of the shins of the little hare.*
> *I will take care not to go towards the dark.*
> *I will go towards the day.*
>
> INUIT SONG[11]

Hot tea was in demand when we arrived at the summer fishing camp of Abraham, a shaman in *Atanarjuat* and grandfather of our knowledgeable guide, Jayson. Huddling inside the cabin next to two camp stoves, we awaited the lard-rich raisin bannock (bread without baking powder) made by our Inuk friend, Aaju, who had come from Iqaluit to join us. Diet-conscious Qallunaat women devoured as much bannock as Aaju could make. Wearing a hand-woven Pangnirtung hat with pins from every Nunavut hamlet, Abraham, the wise elder, arrived on his own skidoo and joined the group. A memorable character, his beguiling smile captivated us all.

After warming a bit, our young guide offered us by turn a ride in his skidoo, to see the first rising of the sun. The land and all its creatures were silent, as a thin slice of luminous flaming red appeared just above the horizon, casting its light onto the mauve and pink of the morning. Our breath made little clouds in the extreme cold; we said nothing. The Arctic year was reborn; the light with its warmth was returning. The cycle of life, so dependent on the sun, was

FIGURE 34 First sun appearing in Igloolik at forty degrees below: traveling the old way by dogsled.

intact. In this unforgiving wonderland, I tried to imagine Inuit using all their skills and knowledge, controlling all their fears, in order to survive.

Just as the Inuit hated to leave the land and its beauty, so we were pried loose from our reveries. We jumped aboard our sleds and headed home, hearing commands that sounded like "oiy" (when we turned right), "Ieee" (left), "wa-wa-ie" (forward), and thankfully "wow" (stop). The dogs obeyed abruptly as well, and we were regurgitated in front of our hotel, tired and cold, but happy.

Knowing some of us were anxious to try char, the great fish of the Arctic, one Inuk responded by retrieving a char from his freezer. He plopped a whole frozen fish on a table in our hotel, and left without a word. Lacking ulus, the woman's knife, we used my jackknife to cut small slices of the pale red fish. The slivers melted in our mouths; tender, with a mild but elegant flavor. I felt like Kazanuuti again, even though the fish that we caught and ate on the Kazan were lake trout.

To celebrate the sun's return, men with torches appeared on the beach. Tradition declared that all lights in the community should be turned off, and the flames of the qulliq extinguished. The resulting dark emphasized the glory of the sun's return. Schoolchildren surrounded the torchbearers, and were full of joy knowing that light and warmth were on the horizon. The solstice celebrations continued in mid-afternoon. A formal written program announced, "On this day the whole community must start a new life," indicating "hope for the future." In the hamlet auditorium, female elders dressed in their amautiit lit two qulliqs. A number of speeches and prayers were given, followed by dancing, juggling, throat singing by Pakak's beautiful wife, and drum dancing by Pakak. The challenge in using the drum (*qilaut*) is to sing while leaning forward with bent knees, swaying slowly to an inner rhythm, alternately hitting the frame of the rotating drum. Periodic cries or exclamations punctuated the song, and sweat dropped off his brow. Pakak later gave lessons to members of our group, one of whom seemed transformed by the experience. During the celebration, our group mingled with the Inuit. One member, a teacher and talented artist, gained the attention of ten to twelve young children whom she was drawing. No child was shocked by her bright blue hair: she simply sat on the floor, legs extended in Inuit style, with a sketchbook on her lap, surrounded by happy faces.

In honor of the return of the sun, several elders constructed an igloo on the hillside above the town proper. All of us relished a visit, but none of us slept overnight. Rowley described his reaction on first entering an igloo: it was like "discovering a new world."[12] Our igloo was warm and cozy even though it was −40°F or lower outside. Caribou and other furs were haphazardly spread on the raised bed platform. The inner wall was plastered with newspapers to prevent cold air from entering. Susan Avingaq, a well-known Inuit seamstress,[13] tended a qulliq. Numerous visitors gathered around to see how she tapped the moss wick gently in the seal-oil-filled soapstone vessel. These traditional "stoves" were used for warmth and light, as well as cooking and heating tea water. Susan's attractive daughter, Annabella Piugattok, star

FIGURE 35 Unknown, *Kneeling Woman Tending Qulliq*, 1951, Inukjuak. Reminds us of our friends in Baker Lake and Igloolik, although the qulliq often now is a Coleman stove.

of the movie *Snow Walker*, showed us Susan's carved doll heads, which were yet to be fitted with proper clothes. They were to be given to relatives and close friends, a disappointment for the collectors among us.

There are artists in Igloolik and nearby Hall Beach, some of whom are known internationally.[14] They are carrying on a very old tradition. In an article by McDougall, Diamond Jenness observed that "such unusual artistic talent among a...nomadic race of hunters could not fail to attract the notice of scientists and various writers."[15] He was writing about the carved ivory art of the Dorset people, much of whose work was discovered near Igloolik.[16] The only item that I took home was lovely pinkish caribou-bone snow goggles, shaped expertly to fit the face, with narrow slits for vision, once heavily used to prevent snow blindness.

Sunday was a day for church—Catholic or Anglican. The Anglican minister's sermon urged the members of the congregation to develop and use their own unique talents. As in Pang and Baker Lake, I thrilled to the music of the old hymns that had traveled far from my Rhode Island Baptist church. But it was not time for reflections, for plans were being made for departure. After a final luncheon of char chowder at the older Inuit hotel, we walked up the hillside to the cemetery outside town with Jayson, our young guide and protector from polar bears. We saw no bears, yet a hint of danger permeated the extreme quiet. Three enormous inuksuit stood majestically above the town as a welcoming marker for those seeking the warmth and safety of Igloolik.

There was time for reflection on return to Ottawa. Images of the natural world lingered: the vastness of the landscape; the invasion of my body by bitterly cold northern winds; the crisp, clear, clean air as it filled the lungs; the pervasive moon as we waited for the sun to return; the pastel colors of the sky surrounding the stark red brilliance of that newly emerged sun; the silence, perturbed only by the sound of snow underfoot, paw, or runner. Traveling with a group of women who had much creative energy, albeit within a tour format, stimulated further contemplation. In Ottawa, therefore, I revisited the magnificent tea-party sculptures of suffragette women on the hillside next to the Canadian Parliament buildings. The sculptures honored the women who paved the way for all of us northern and southern women. Curiously, we had not discussed our roles compared to those of the Inuit women we met.

Over the ensuing years, however, through email correspondence with Jeena, I was to learn more of the life of the modern Inuk woman. In response to a book we later sent her titled *Inuit Women*,[17] Jeena revealed, "It is interesting how men treated women back then. All in all it is great how we new generations of women are being treated, not like our parents and grandparents." Her friendship has fostered a reciprocal, learning, and enduring adventure between members of two cultures, unique in their traditions and lifestyles, but similar in their members' personal hopes and aspirations. Jeena would prove the stimulus for return visits to this still-isolated, but now electronically accessible Inuit community.

10

TUNDRA CAMP: BAKER LAKE

We women were treated as outcasts even though we were ordinary people.... We used to drink water moments before the sun went down, because we could not drink liquid or eat anything until the sun came up.... We certainly did have strict taboos.... We women were always out hunting, sometimes saving our own husbands.... We used to drive dog teams.... Oh, we used to be in such desperation and now when we look back we can laugh about it.

ELIZABETH NUTARALUK[1]

AUGUST 2007 Soft evening light peeked through thin areas in Annie's hand-sewn caribou-skin tent. Reclining in our sleeping bags on caribou skins laid on the gravel, we were only vaguely aware of the slightly gamey animal odors. The air was cool, and we were feeling mellow. We were excited to be back on the tundra, and living with an Inuit family. Midway through the night, the camp dog started barking, and Joyce was sure a wolf was on the prowl. She grabbed her bear whistle and flashlight to face whatever demons were there, while Fred snored away calmly, unaware of imminent danger. Joyce's preparedness and Fred's strength even in sleep must have been read by the "wolf," for the dog, and we, were alive in the morning.

Sleeping in that tent was just one of the reasons we wanted to visit Baker Lake again. On the 2005 canoe trip down the Kazan, two years before, we were trapped by a gale at the bottom of the portage around Kazan falls, and never saw the last thirty miles of the river. We planned to do the Kazan again, starting at a different entry point. The archaeological sites are rich on the lower river, and we could once again visit friends in Baker Lake. Rob Currie was eager to go again, too. We found another couple who wanted to do the trip. We kept working in the gym.

It was not to be. In April Fred developed sudden dramatic weakness in his right arm, a serious problem for anyone, particularly a canoeist. Ever an optimist and in full denial, Fred nevertheless purchased expensive Calm Air tickets to Baker Lake, thinking he would improve. Careful medical evaluations proved there was a cervical spine disc problem. In June we gave in to the inevitable, telling our partners it was no go. Conservative management eventually resulted in full recovery, but it was too late.

Our guide, Rob, mentioned that Annie, a friend of his who sewed wall hangings, made a caribou-skin tent the year before. Hope for a different kind of visit emerged, one where we might live with an Inuit family and begin to learn more about the life of the people, now and in the past. We called Annie.

"You can stay with us in our home, and we will go to our camp too, but the home will be crowded.... Our daughter, son-in-law, and child are living here.... There are not enough homes. But the tent is plenty big." Annie complained about the very cold spring, with almost no thaw on the lake or the Thelon at the end of June. "Maybe there will be no mosquitoes this year."

Her offer rejuvenated us. We would go to the North again.

As usual, we flew through Winnipeg to Churchill, Arviat, and Rankin, arriving at Baker Lake about 11 p.m. On the flight over the Barren Grounds, we were full of nervous anticipation, not knowing exactly what to expect from our hosts in Baker Lake. Annie greeted us at the airport. She recognized us easily enough, because we were the only elderly Qallunaat couple among the throngs in the little lobby; we could see in her smiling face that she was to be our hostess. She looked friendly and entirely welcoming, and we felt at home at once. Her ATV could not possibly carry three adults and lots of fishing and camping gear, so we piled into an old cab and followed her home. A couple of ATVs and a temporarily disabled Ford truck were parked in front of her house, and two skidoos that looked in excellent repair were to the side. Entering through a covered vestibule for taking off boots and outside gear, we met her husband, Joe, who tried valiantly to cross the lake and get to us in his twenty-foot-boat two years before, only to be driven back by the gale.

Susan, an adopted daughter, lived in the home with her boyfriend and their eighteen-month-old daughter. Despite her runny nose, the little girl was the object of everyone's attention and affection. Susan's boyfriend was from another hamlet, and he, Susan, and their child intended to return there very soon, no doubt after they had helped Annie with the cooking during our stay. We were imposing new burdens, but the family rallied around Annie. One of her sons was playing guitar in his back room, singing gospel songs. An adopted teenage daughter was at a high school dance; at Annie's instruction, we plopped our sleeping bags on top of her bed, worrying about where she would sleep after she returned from the dance. We were intrusive, yet everyone was very kind to us.

It was uncertain when we should go out to their camp, and what we should do while in town. Annie's regular responsibility was leading Sunday church services. Both Annie and Joe seemed pleased when we suggested we stay in town for two days, which allowed us to visit friends, and to go with them to church.

Eva, the sewer of a favorite wall hanging in our collection, was a person we wanted to meet. She is the birth daughter of our friends Ruth and the late Josiah Nuilaalik, and sister of our friend Hannah, but was adopted out to grow up in the home of Ruth's stepmother, Marion Tuu'luq, and father, Luke Anquhadluq. A short walk took us to Eva's home. Her husband Paul spoke English and interpreted for her. Seven kids emerged from corners of the house. We learned that it took Eva eight to nine months to complete our wall

FIGURE 36 Eva Ikinilik Nagyoulik, *Untitled*, 2006, Baker Lake. Wall hanging depicting, from above, the rivers and lakes, a rainbow, and the wide caribou tracks extending across the tundra. The image consists entirely of embroidery floss stitching.

FIGURE 37 The tundra in fall below Jessie Oonark's grave, Baker Lake.

hanging, a five-by-three-foot abstraction of the tundra, done entirely in bright pastel embroidery floss in detailed feather stitches—painting with a needle. Paul said he made her take out the initial stitching when he saw it went all the way through the wool stroud; the only way to sew these, he said, was to make sure the stitches do not go all the way through, as in the old days when it was essential to keep clothing waterproof. He learned this from his mother-in-law, Marion Tuu'luq, who lived with them until she died. They confirmed that the scene shows the tundra with flowers and lakes, caribou tracks, and a rainbow, all in glorious, vivid color. The perspective is primarily from above, as in flying over the tundra. Only now, in Eva's middle age, was she carrying on the wall hanging tradition of her famous mother and grandmother. Perhaps having seven children was sufficient explanation for this late start.

Paul and Eva sent two of their daughters to guide us to Hannah's house. Hanna's daughter recognized us at once, excitedly informing her mom that Joyce and Fred had returned. Hanna became quite animated and full of stories about her present life, including her pleasure in visiting friends on the outside, in Winnipeg and elsewhere. She spoke proudly of receiving a firefighting certificate from a school in Rankin. An old letter from us was attached to the door of the refrigerator; we had been friends for four years. Later she took us over to her mother Ruth's home for a reunion.

Visiting our friends in the Nuilaalik home was a highlight in many ways. Ruth recognized us immediately, pantomiming canoeing and then dancing, telling us without words that she remembered us. Two years had elapsed since Josiah's passing; she was adapting and seemed much more alert and happy. Ruth has a truly delightful affect, and a very expressive face. She had just returned from a family excursion by charter flight up north to the old family camps on the Back River. One of her sons (there were fourteen children, nine surviving) arranged and paid for this trip; he apparently has been very successful in the modern commercial world. They caught a lot of big char (*iqaluk*), but Hanna told us Ruth will never go up to the Back River again. Living on the land was too much work, and there were too many painful memories, too much history. That was almost fifty years ago, but we surmised that the agonies of the starvations remained very fresh for her.

On a beautiful day under billowing clouds, Joe drove his boat about twenty miles up the Thelon, past rolling hills and some cliffs, through several strong rapids, seeing one distant musk-ox. Using "daredevils" with casting rods, Fred and Joe caught a few trout, one about twelve to fourteen pounds, in only a few minutes. Long strips of succulent red eggs (roe) were quickly removed from the largest fish, and Annie and Joe clearly relished eating it raw. We also snacked on strips of gray and very dry caribou (*nikku*). The dried caribou was a bit of an acquired taste to us elders from a culture more attuned to fresh meat from the supermarket. The caribou was from their own larder, shot on the tundra and dried and stored at their camp.

Back home, we were treated to a delicious, juicy roast caribou dinner. Annie insisted we sit at table, possibly because she noted that we are just too stiff to sit as they do—on the floor with legs straight out in front. Everyone else sat on the floor close to the caribou, which was on a piece of cardboard. Annie was expert with her ulu, carving the caribou with a swift, sure rocking motion. Friends came in, finishing the leftover meat, departing after a short visit. The next night, dinner was succulent, fresh moose that had been curing on the ground outside for several days, provided by one of Annie's sons. Which is better, moose or caribou? Perhaps moose, but it is a close call.

Sunday is an important day in this family, going to church for two hours in the morning and an hour and a half in the evening. We joined them for both services, and were greatly impressed by the strength of their belief and by the friendly manner with which we were greeted. Attendees were mainly elders, but the church was almost full. A handsome Inuk Anglican priest welcomed us in English, and everyone clapped. The rest of the service was in Inuktitut. Annie is a deacon or its equivalent and gave communion to almost everyone, including us. Annie wore her leadership role in the church easily; a calm, kindly, and self-assured presence, very dignified in her white robe. People offered to share their hymnals and Bibles, which we appreciated, but they were in Inuktitut

FIGURE 38 Ennutsiak, *Bible Study*, 1955, Iqaluit. Preaching and singing the word of God, a powerful force among the Inuit.

and were not very useful to us. Evening services were less formal, with many testimonials, and lots of emotion. An occasional person fell down, crying and in great distress. Annie joined one of the parishioners, clearly giving comfort to the one in need. Tuesday services were for youth and were said to be full of intense and joyful singing, clapping, shouting, and dancing. We observed this later in home camcorder movies, played on their television set.

There was plenty of time for just talking, sharing stories, and getting to know one another. Annie was from the upper Kazan region, and grew up south of Baker Lake among the Ahiarmiut. Joe was Utkuhiksalingmiut, from the Garry Lake–Back River–Chantrey Inlet area north of Baker Lake. Annie's relatives included the famous Arviat artists Luke Anowtalik, who

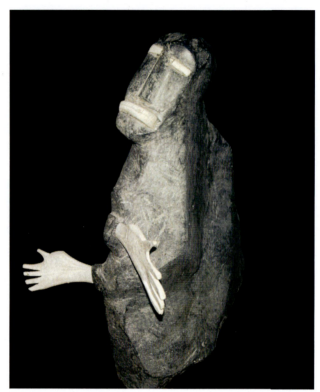

FIGURE 39 Elizabeth Nutaraluk, *Female Shaman*, c. 1982, Arviat. Imploring the powers above.

was one of her uncles, and Elizabeth Nutaraluk, who was one of her grandmothers. We had carvings by each of them, including one of our favorite pieces, a very large, rough, gray stone and caribou antler piece by Nutaraluk that was carved on the tundra outside of Arviat, depicting a female shaman or an anguished woman either imploring the gods for help or bemoaning her fate. Joe and Annie smiled appreciatively on seeing pictures of Anowtalik and Nutaraluk in the 1950s on Ennadai Lake in a caribou-skin tent and kayak in a book that we brought with us.[2]

Annie's fists were clenched as she talked about rivalries experienced among the various groups that moved from the land to Baker Lake in the old days, about fifty years before. The early comers spoke English and had better jobs, and lorded it over those who came later, including Annie's and Joe's people. This story was the same that Elizabeth Kotelewetz related in our first visit to Baker Lake in 2003. Now the Ahiarmiut and Utkuhiksalingmiut are learning to speak up and assert themselves. People still tend to stay within their own groups unless the community is threatened, in which case all come together.

We spoke of our four children and ten grandchildren, a small family by Inuit standards. Their family structure was more complicated, with adoptions out and adoptions in, typical of Inuit culture. Their youngest son was adopted as an infant. A son was adopted out to Annie's mom, but he froze to death at age 21. Another son was adopted out as a small child to her stepmother, and he remained a regular part of the family circle. Both of the daughters in the home were adopted. One natural-borne daughter died at age five of congenital heart disease, and another was adopted out; she was soon to be married, but Annie was not going to the wedding, relying instead on pictures of the wedding. Adoption plays a crucial role here; every child is valued, and protected.

Annie worked as a cook in the elders' home. Joe and their eldest at-home son also had stable jobs with the town, driving the water and also the sewage disposal trucks, the only way here to deliver water and dispose of human waste. Mines were being developed: gold was discovered north of Baker Lake, big diamond mines were located only one hundred miles to the northwest, and a new uranium mine was being planned west of town. Joe was worried about the long-term effects of the mines on the environment, but people needed work. He would rather expand his occasional boat business with his sons, picking up and delivering people to the Kazan or up the Thelon as far as Beverly Lake, one hundred miles up the river. Joe spoke about the changing weather in a calm and unemotional manner, but we inferred he was resigned to the situation. He admitted the weather was changing; there was less snow, and less water in the river; the ice was thinner, and the snow was not very good for making igloos anymore.

Annie and Joe had two camps: one six miles out and another about sixty miles up the Thelon. We were to visit the closer of the two. As we were mounting Annie's ATV, we ran into three kayakers passing through town on a great voyage. They started the year at Great Slave Lake, taking the famous nine-mile portage (Pike's portage) to Artillery Lake, and then down the Hanbury and Thelon rivers to Baker Lake. Their season would end at Chesterfield Inlet, where everything would be stowed for a final paddle up to Repulse Bay and, eventually, Coppermine. They looked serene as they paddled down the lake in two kayaks, a double and a single. As envious as we were of their trip, they looked envious of our laughing, friendly relationship with Annie as we climbed aboard the rear of her ATV, sitting on a caribou-skin padded rear seat for the drive to camp on the tundra. It was a tight fit, with Joyce hanging on to Annie's ample girth, and Fred trying not to fall off the rear.

A host of small plants decorated the countryside with their blooms and colorful red leaves: mountain avens, Labrador tea, saxifrage, dwarf Arctic fireweed, Arctic cotton, lingonberries, blueberries, cloudberries, and crow berries.[3] A small, low plywood camp building and a very small outhouse were nestled on a plateau above and perhaps a fourth of a mile from the Thelon. Annie's beautiful, handmade caribou-skin tent dominated the scene. The skins were a symphony of color, very rich in tone. Old caribou skins and bones were scattered on the tundra near camp, and a single musk-ox skin was on the ground; burned caribou bones were piled deeply in an old oil barrel. A large box contained dried caribou meat. A barrel contained water that was brought to camp from a small stream behind camp, or on occasion from the Thelon. The roof and door of the cabin were low, bruising Fred's head several times upon entering. The usual exclamations were suppressed. A woodstove used willow and birch twigs for fuel. A caribou skin and an unfinished work on cloth (a

FIGURE 40 Labrador tea—ubiquitous and intensely fragrant.

planned gift for Rob) were draped over an old couch. Another unfinished wall hanging was in an adjacent room. We pledged secrecy so that the wall hanging would remain a surprise for Rob.

Our meals included a lot of country food: moose meat that we brought with us, golden-red lake trout that we caught in the Thelon, blueberries and cloudberries picked off the land, and bannock. The food was delicious, although dried caribou was tough even after it was pounded with a hammer on a rock to soften it up.

The tent was made from twenty-five caribou skins harvested last fall by Annie and her family, sewn together with tight caribou sinew stitches under the direction and with the help of one of her aunts. It appeared identical to those photographed by early explorers.[4] Standing about ten feet high, we could stand up in the middle. The poles on which the skins were hung were the hardest thing to obtain for they had to be brought in from hundreds of miles away. A ring of boulders held down the tent's outside edges, and a thick flap allowed entry and exit. Most of the skins were thick and covered with fur in various shades of brown and cream, but a few thin areas let in a bit of light. This was Annie's first tent, and she could not hide her pride in it. Indeed, it was a work of art, soft and rich in color, varied in hue. We delighted in being able to stay in it. Annie stayed in the cabin, respecting our privacy, although in the old days we certainly all would have been comfortable sharing the spacious tent. She insisted we bring a large caribou-skin rug from the cabin and also a foam mattress to cover the gravel base, and also that we take a caribou skin home with us. The caribou skin stayed there, where it is much more useful than in the heat of North Carolina.

From the tent there was a long view over Baker Lake to the east looking all the way down to where the Kazan enters on the south side. The Thelon flowed below camp, with low hills on the opposite bank. Annie saw a white wolf recently, and had shot several wolves here over the years. The wolves probably were drawn to this area because it was the site of an annual caribou hunt. A few flowers were in bloom around camp, mainly Arctic fireweed.

The tundra was dry in most places, but there were abundant low blueberry bushes in low boggy areas, just ripening. We scared up a number of big Arctic hares in summer gray and brown colors, especially on top of the rocky hill above camp, and saw a raven. A pair of sandhill cranes flew by, and a nesting long-tailed jaeger rested on a stony nest. Annie loves the peace of the tundra, and often spends time here alone.

Nearby there were remnants of three-thousand- to four-thousand-year-old pre-Dorset ("Paleoeskimo") camps, and a bit farther along there were several old Thule campsites from around five hundred years ago. This is a spot where the caribou cross the Thelon. Caribou congregate nearby ten months of the year, many staying through the winter. The caribou would return the following month, but we saw not a single one, despite Annie's repeated searching of the horizon with a spotting scope.

With just the three of us, there was a lot of talk about current conditions, the old life, people, and family. Annie told many stories of the old days, and we were eager to hear them. She grew up on the upper Kazan, and for a time lived near Yathkyed Lake, where we started our 2005 Kazan River canoe trip. She once lived on Ennadai Lake, and later near Dubawnt Lake, which sites nearly perfectly describe the limits of the territory of her people, the Ahiarmiut.[5] She recalled a time in the spring of early childhood when there was a long starvation, when one of her grandmothers died next to her in the tent. A grandfather, his second wife, and another young woman also died, "just in our camp. We were eating the skin tent to try to stay alive.... Mother and Father were so weak they could not walk. Finally caribou were seen, and my father got up and tried to walk to them, far away. He fell down and stayed down, and my mother followed, lying down next to him. Were they dead? Was my dad dead? They just stayed there. Maybe they were crying. They wanted to feed us." Finally an uncle appeared over the hill, traveling with dogs from another camp. "He shot a caribou, shared the meat with us, and saved our lives.... We started by eating small bits only."

Joe also grew up on the tundra, and like Annie, was evacuated as a child to Baker Lake, where they met. After their marriage, they spent about half of each year for ten years living the old life among Joe's family up north on the Back River and Chantrey Inlet. They stayed in skin or canvas tents and igloos, trapped foxes with dog teams, ate caribou and moose and char. "It was hard...an adjustment.... I was just out of school...not used to caribou skins...always eating caribou, caribou."

The community has a number of social problems, and we discussed some of them. It is sad that some people commit useless crimes and violence, and are sent far away to prison. Soon Annie tired of such painful subjects and tried to teach us some Inuktitut. We all loved the instruction, and friendly laughs filled the cabin. Once Fred seemed to get the pronunciation of a word just right, which delighted her. Despite the stories of difficult times and social problems, Annie was happy, gracious, warm, and positive. Her energy and good cheer were infectious. She hugged Joyce, and probably wanted to hug Fred for his rapid learning of one word of Inuktitut.

There were many more stories. A woman froze to death last year in town walking to the Northern store. Two of Annie's sons got lost on a skidoo, ran out of gas in a storm, failing to come home. A search started, and friends came with food to stay with her (Joe was away). An uncle drove out searching for them and saw a beam of light from the heavens, over a lead in the ice, and there they were!

Annie recalled her "rescue" by airplane many years ago, when six children dressed in caribou-skin clothes were taken from an igloo on the tundra to Baker Lake and put into school. Other relatives were taken to Arviat. They were taught the ABCs, were not allowed to speak Inuktitut. Eventually, their parents migrated into Baker Lake to be close to their children. For these misdeeds, the government "has apologized and has agreed to pay a big amount to us and others like us for life." A raven landed on the cabin and called, catching Annie's attention. She was quiet, and then said, "I should not be angry."

The wind came up suddenly one afternoon, so strong as to make standing difficult. The caribou tent was blown out at the bottom. We worked together, turning it so the door did not face the wind. Annie insisted we move into the cabin for the night. The rains came, and the cabin leaked prodigiously from every corner, requiring placement of every available pot and can and bucket under the steady drips. The flat roof needed repairs, even though they replaced the tar paper on the roof just the year prior. We were forced out of the bedroom and slept in the relatively drier central room. Could the tent have been wetter?

On a bright and warming day we picked ripening, plump, juicy golden cloudberries in a peaty boggy patch. We climbed a hill overlooking the tundra and the lake, and lay down in the warm sun, where Fred drifted off with dreams of boyhood. Why do we love the silent moments so much? The Inuit have a word that describes how we felt: *quvianaktuq*, to feel deeply happy, a feeling we experienced many times in the North, especially when we were alone or very quiet. Contentment is more accessible here, away from the crush of our everyday world.

Annie showed us how to start fire for tea with dried moss. She said caribou were at Aberdeen Lake now, well up the Thelon. Vera gave us a lovely name: "elders who seek understanding or who are wise." She wrote some words for us: *eleesimayuk* (wisdom, knowledge); *tuqiseeayuk* (understanding), *matna* (thank you); *eyounaqtuq* (funny); *inutuqak eyounaqtuq* (elders who are funny); *piqanara, elanara* (friend). Elders who are funny, not a bad approbation, at least as good as elders who are wise. We felt close. We are all the same—except they may be more generous, more caring, and certainly less materialistic than we.

Breakfast on the last morning in camp was Joyce's treat, a blueberry bannock, which turned out to be a gooey mess, but tasty. Susan and her boyfriend came to camp again on an ATV to help carry our gear back to town, with her cute little daughter in her amautik. Back in town, we retrieved the warm shawl Joyce had brought for Annie. Annie was moved. "I had a good time in camp.... You were so funny speaking Inuktitut.... It was good to be there with someone."

Annie gave us a small piece she sculpted from black basaltic rock, a musk-ox with a human face. We did not know she carved. She is talented in many ways; able to survive on the tundra, sew caribou-skin tents and clothing, hunt and fish, but also adept on computers and digital cameras, seemingly equally at ease in the old and the new worlds. Talk centered on hopes for the future, theirs and ours. There was much sadness in parting; Annie's eyes looked moist, as were ours. It is a long way to get there; we might never see them again, and we cared about them very much.

At the airport we overheard a tattooed white man talking loudly to a large, handsome Asian man in brand-new camouflage pants with a computer bag slung over his shoulder. The tattooed worker was relating his experiences in the mining camp north of town, where he was hit in the shoulder by an amazingly strong Inuk, knocking him across the room. He was afraid of retribution by the husband of an Inuk woman who worked in the camp and "serviced" him. Clearly these miners had neither understanding of nor respect for the Inuit. We worried about the future, and hoped that the Inuit do not lose their place in their own community as outsiders pour into town, seeking gold and uranium.

In December, we talked again by phone, and learned that Susan moved back to her boyfriend's hamlet just after we left. Annie did not get her skin tent to Rankin for the regional craft exhibition, but at least she had christened her caribou tent with visitors who loved the opportunity to be Inuit for several nights. Two groups of caribou came in large numbers in September. One herd was "not our caribou; they walked through." The other caribou stayed, were still there three months later in the depths of winter, and some were sleeping in town relatively close to their dogs, near their home. Annie did not seem surprised by this. They had not and would not shoot the caribou that were staying in town.

Although we couldn't see the scene, we imagined it: The hamlet is blanketed in snow, virginal in its purity, abutting the lake deeply frozen and covered in white. The pale pink gravel of the main road is covered in drifted snow. The sun appears low over the horizon for a few hours a day, and the colors are those of dusk, with a rosy glow on the margins. Quite shockingly, a small herd of caribou rests peacefully near some homes on the edge of town, up near the snow fence, not far from huskies tied to stakes near their doghouses. Everything is quiet. Perhaps the caribou are escaping harassment from wolves outside town on the frozen tundra. People wander outside at times, in their parkas or amautiit, some lined with wolf fur around the front of the capacious hoods because wolf fur is resistant to freezing when wet. It is $-35°F$ ($-37°C$). People and animals are hunkered down, intent on surviving the cold, coexisting in a manner that suggests a truly peaceful kingdom. We see the community engulfed in snow as clearly as Eva saw the tundra from above when she made her wall hanging of the Barren Grounds in the bright pastels of the Arctic summer.

11

AUGUST ICE: IGLOOLIK

That's my prescription for each of us. To be driven into action by the wild beauty and difficulty of a place; to make decisions about it based on biological health—what we can do for the earth— not how much money we can pimp from it.

GRETEL EHRLICH[1]

AUGUST 2007 Jeena wrote Joyce in early 2007: "My boyfriend does not think I will ever see you again.... When are you coming up here to Igloolik?" Much was shared over the Internet: how Jeena was doing in school, how Joyce was adapting to retirement, details of each one's family, books read, distress about too much or too little ice, about deer proliferating and caribou decreasing. Joyce wanted to introduce Fred to Jeena, and now Jeena more or less invited us. We discussed how we might get to Igloolik from Baker Lake, extending our journey following our visit with Annie's family. The only way to do it was to fly east from Baker Lake to Rankin Inlet, and then across Hudson Bay to Iqaluit. An overnight stay would be required before we could catch a direct flight northwest over Baffin Island and Foxe Basin to Igloolik: two sides of a large triangle, but the only available commercial pathway. Discussion shifted from how we could do it to what we would do once we got to Igloolik.

A lengthy correspondence ensued. Jeena asked whether we were members of Greenpeace. "We don't like Greenpeace here.... We hunt seals." She suggested we might visit her family camp out on the tundra, and go with them in their boat to look for walrus as well as seals. Before long, however, Jeena realized that Canadian Ranger duty might take her boyfriend and father away during the time we needed them to see the animals, and to fish for Arctic char. Our concern increased when Jeena did not answer email or phone messages for several weeks. Because her grandfather in Hall Beach died, Jeena and her family had gone there to be with family, and had no access to the Internet. Moreover, her brother recently committed suicide. Jeena was depressed, but was doing better. A backup plan was needed, so Fred contacted the only guide listed on the Internet for Igloolik.

No ice was seen in Hudson Bay on the flights from Baker Lake to Iqaluit. A brief stop in Rankin Inlet was a bit of a disappointment; too commercial for our tastes, and the renowned ceramics studio was closed. The Frobisher Inn in Iqaluit welcomed us for a fourth stay, and we enjoyed a dinner of fresh char on our forty-fourth wedding anniversary. An effusive Qallunaaq waitress was so impressed with the durability of our marriage that she bought a bottle of quite good wine for our dinner, testimony again to the friendliness of the people in the North. She got a really big tip.

Flying to Igloolik, we passed over Baffin Island, and then over massive patches of pan ice on Foxe Basin, marveling at the wild and intimidating spaces below. The Baffin Island coast along the eastern margin of Foxe Basin was among the last to be mapped in the Canadian Arctic, a task finally accomplished in the 1930s by Graham Rowley.[2] Peter Freuchen, Knud Rasmussen's principal colleague on the Fifth Thule Expedition, failed in a similar attempt fifteen years earlier when he badly froze one leg trying to get to Igloolik by dog team in deep winter.[3]

A small patch of dense fog covered the island of Igloolik and prevented our landing. All landings must be done by sight. With limited fuel, we could only try one additional pass. Fog is common here; the same thing happened the day before. Frustrated, we flew all the way back to Iqaluit with a brief stop in Hall Beach, a former defense early-warning (DEW) line radar installation, only forty miles south of Igloolik. With the end of the Cold War, the radar site was abandoned and the towers now serve as a marker for travelers—a huge technological inuksuk.

The second time was the charm: we landed in Igloolik without incident, along with a very well-behaved adolescent hockey team. The ice in Foxe Basin stretched all the way to Hall Beach, but the seas around Igloolik were clear. Jeena and her family were among the throngs in the airport, as were our guide Brad and his Inuk wife. Jeena was short, attractive, and bright-eyed, with an infant child in her amautik, and a young son and her boyfriend at her side. It was awkward with two hosts. Jeena and Brad had a short discussion off to the side, and they quickly determined how they would share custodial duties. We would spend that evening with Jeena at her family's camp, and then would go to Brad's camp on Igloolik Point for the next four to five days while Jeena was at work. The last day or two would be spent with Jeena and her family in town. A bunkhouse usually reserved for the scientific research center was home for the first two nights.

Igloolik is a flat, horseshoe-shaped island, with an eastward arm leading to Igloolik Point. The hamlet is situated at the deepest part of the horseshoe. A great many boats were gathered in the harbor. The ice went out a month ago, and the water would be open for another two to three months. Nineteen months ago on her winter visit, Joyce had seen boats pulled up on shore, covered in snow, and bathed in the dim light of the moon while waiting for the new sun.

According to plan, we met Jeena at the harbor. She was dressed in her purple amautik with small child inside, accompanied by her two other young sons, two of her brothers, and her lean and handsome boyfriend, Scott. The

FIGURE 41 Scott, Jeena, and their son by Foxe Basin.

family boat had twin four-stroke engines, and a plywood cabin tethered to the frame of the boat. Food and water were loaded, and we cruised to their camp on Melville Peninsula, some fifteen miles distant. Along the way, Jeena told us that she was the only one of her siblings to finish high school. We knew she was in the Arctic College studying to be a teacher or maybe a nurse. Scott and her brothers had intermittent work of various sorts, but really were most interested in being hunters. Her father and mother had accepted Scott as a leader, and he was captain for this trip. As we approached their camp, we surprised a pair of elderly Inuit, who were emptying the char from Jeena's brother's nets he had set the previous day. Our friends were unconcerned, and everyone waved cheerfully: "It is OK.... We do not worry.... They can have them.... There will be more char in the nets soon."

Camp was a plywood cabin with an outbuilding and a partially finished second cabin. About twenty large, beautiful, red-fleshed and silver-skinned char were drying on a rack, covered with a plastic tarp to keep birds off. The gravel and stone beach was almost entirely devoid of plant life; it reminded us of Beechey Island. Many fossils showed on the exposed rocks, and there was a pile of collected fossils for looking, but never to be removed by decree of Jeena's mother. The grounds were entirely clean of cigarette butts or any other debris, by a second decree. One caribou skin was drying on the ground. The broad gravel terrace was flat; green tundra and low hills were visible well behind the camp. There were no other camps, and no other people. A lonely expanse of water and wide-open space dominated the scene.

The little cabin was tidy, with a small main room containing a kitchen and table, and two small rooms with sleeping platforms. Supper was dried char, frozen caribou sliced off in thin strips with an ulu and eaten raw, boiled caribou, Cokes, cookies, and tea. The cabin roof was very low; the baby thrown overhead with much laughter was only our concern. Small talk occupied us; we were happy to be sharing this meal with them. It cost a large carbon footprint and more than a few dollars to get here, but that was far from our minds.

The nets were inspected by the fishing-expert brother, but were empty. The skies started to turn pink, and all too soon we started the return trip. Jeena had to be back at her job the next day. But adventure was lurking nearby; seal heads started appearing, easily spotted in the calm seas. Hunting seals is a rhythmical dance with family members participating: The seal pops its head through the glistening sea; it is recognized by the sharpest eye; the boat captain moves rapidly toward the spot; the hunter steadies and fires his single-shot gun; the boat moves swiftly to the stricken seal, where it is harpooned before it sinks.

So it was with our team. After the shot, Scott revved the motor and raced to the seal; the harpoon was in the brother's hand and then in the seal, directed with precision. Neither of us saw water placed in the mouth of the just-killed seal. Once it was customary to give thanks to the seal by placing water in its mouth, easing the passage of the seal into the next world. Jeena offered, "We don't do that anymore." After tying the seal to the side of the boat, we rode slowly toward the shore, dragging our catch and leaving a stream of crimson in our wake. It was a young bearded seal, still immature

FIGURE 42 Char hanging to dry as we arrived at Jeena's family camp on Foxe Basin.

FIGURE 43 Janet Kigusiak, *Char Hanging to Dry*, 1999, Baker Lake. A vibrant abstraction of our favorite arctic fish using tissue paper and acrylic polymer.

although it was about seven feet long. Three men were required to drag it onto the shore.

Drawn by the sound of the gunshots, another boat appeared, and its hunters joined in cleaning the seal. The glint of blades brought us to rapt attention. The knives were wielded expertly. Skin and blubber were removed in foot-wide strips, and set aside for use in making soles of kamiks and whips. Meat was prepared in chunks. Skin and meat were shared equally, but the intestines were treated differently. Intestines were cut into sections, cleaned rapidly with a stripping motion, braided and secured on our boat. "The best part to eat, the intestines," they all agreed.

The sun was setting; the skies turned a rosy red color, outlining the hunters bent at the waist with their long knives eviscerating the carcass of the seal. We imagined the same scene occurring over the centuries, little changed except for the type of boats and motors used to get here. Coats were drawn tight against increasing cold, and wool caps pulled down. The water was purple at our feet, stained in blood. Half of the sky was bathed in an intense rose, pink, and orange glow, reflecting off the calm waters. The light penetrated and echoes still.

Midnight light was fading to a dull gray when we arrived in town. All of us walked up the dusty gravel streets to Jeena's parents' home, not far distant; the seal parts were brought by ATV. The entranceway displayed skulls of a polar bear and a walrus, as well as a caribou antler rack. The home was crowded, warm and friendly, and was filled with adults and little people. There was much happy talk and laughter. Each of Jeena's parents held steady jobs in the hamlet, and the home showed evidence of their relative prosperity. Jeena's first child, a four-year-old boy whom she adopted out to her mother at birth, took a bottle from his grandmother, while Jeena nursed the infant. Everyone spoke English, flavored by Inuktitut, which was still the first language. The treasured seal intestines were added to a large pot of vegetable soup, meat, and blubber, but we southern elders became exhausted before it was finished cooking and returned to our beds shortly before

2 a.m., missing what undoubtedly was a special treat. During the period of twenty-four-hour light, the days have little formal ending and beginning for the Inuit. Unfortunately, our internal clocks were set differently.

Jeena wrote to us later that her four-year-old, whom we saw taking a bottle, shot his first seal just after we departed. Within a year he also shot a small polar bear. Photos accompanied the latter deed, along with explanations that the bear was sick, and permission was granted to harvest it.

After years on the land as a wildlife officer in Resolute and Arctic Bay, our hired guide Brad was experienced. His camp on Igloolik Point was perched on a low gravel shore above a wide, shallow bay facing down into Foxe Basin. Camp was quite simple, but had a generator for a few hours of electricity if needed. Brad ran camp and cooked. The dining room was a large plywood cabin with a Coleman stove and a few tables with benches. "His" and "Hers" single-holers out on the gravel ridge served some of our fundamental needs. We slept in a tiny hut, in sleeping bags on an old spring and mattress. Food was very good, especially the fresh char smothered in onions that we caught and provided for virtually every dinner. Other camps lined the same shores, although they appeared empty. A few lonely, pale yellow Arctic poppies stood alone on the pebbled shore, big luminescent blooms standing proudly on narrow stems. Ringed seal heads popped up in the bay intermittently.

Brad knew the land, the water, and the animals, and was a repository of much lore about the North. Problems with Inuit society and behavior, governments, regulations, local health care, and local schooling occupied much of his mealtime conversation. He was suspicious of our note taking, and let us know that he hated people who write books after being up North for brief periods: "If you haven't lived at least one full year here, you don't know anything." He had a point. We continued our note taking nonetheless. He took excellent care of us, the only visitors to the camp that week. One professional photographic group would be coming soon.

We were there between seasons, following the June-July bowhead whale watching season, and prior to the best walrus season. Char, however, were abundant. Brad said char were very tough to catch on a fly, but he knew a fine place to catch them in coves fifty miles north of camp on Baffin Island. We made a pair of trips up to Baffin, and also a pair of trips south to the ice in search of walrus and polar bears. Brad's Inuk wife, their son, and Inuk guide Manassie accompanied us everywhere in a twenty-five-foot boat, with a single, relatively new 150-horsepower four-stroke engine.

On the way north, we either passed to the west of Igloolik, close to the shores of Abverdjar, or to the east, where the waters were open but very shallow and occasional reefs were hazardous. Father Bazin and later Graham Rowley uncovered a treasure trove of Dorset cultural artifacts on Abverdjar, including ivory carvings, tools, and implements, documenting the long use of these lands by Eskimoan people.[4] Included in the discoveries was a beautiful little ivory carving of a flying bear-shaman, testimony to spiritual beliefs apparently similar to those of the modern Inuit before the coming of the missionaries.[4,5] Some of the Dorset carvings had a weird and almost frightening aspect, including what appeared to be tortured or agonized human faces carved onto human scapulas.[5] The Dorset people were abundant around Foxe Basin, and left thousands of old archaeological sites scattered across the tundra, still being investigated by a small team of archaeologists based in Igloolik.

The route by Abverdjar took us along the rocky coast of the northern tip of Melville Peninsula, part of mainland Canada. Islands dotted the waters, and land was usually in sight somewhere. We stopped at Marble Island, where the white rocks were dense with nesting Arctic terns, greater eider ducks, and black guillemots. The water was turquoise, looking almost tropical. Manassie was tempted to eat some eggs, but they were too far along, and were put back on their nest.

Fury and Hecla Strait is notorious for ice and big currents, and hunters sometimes failed to return from trips into these treacherous waters. The strait forms a narrow pass between Baffin Island to the north, and the tip of Melville Peninsula to the south. The strait was named for Parry's and Lyon's two boats that overwintered here in 1822, the first contact of the Iglumiut with white people.[6] A shaman put a curse on the Qallunaat, which seemed to work because none returned for almost one hundred years. We could see deep into the strait, extending far to the west. Little ice was evident now, being dammed up further west and north, but Manassie told us it will come later. "When the ice comes down, the narwhals will come."

Narwhals arrive in herds in late summer or early fall. Bowhead whales also are regular visitors, coming through early in the summer, when thaws open leads in the ice. Our friends claimed to have seen seventy bowheads this year, including a couple of dozen calves, a very large number given their near-extermination in the eastern Arctic, and their rarity as little as twenty-five years ago. Jeena and her family told us virtually the same stories. Brad's wife's tales about the bowhead were vivid. Just a month before, a pair of bowheads came to rest under their boat, tails on one side and heads on the other. One of them scraped its back on the underside of the boat, presumably fixing an itchy spot. Her fright was evident as she told the tale, her eyes wide. Usually the bowheads pass through, with only a few stragglers remaining for the summer. Despite repeated searches with binoculars, we saw no bowheads, only ringed, bearded, and occasional harp seals, and eider ducks.

Approaching Baffin, high, rocky, and still-ice-covered cliffs were crowded with hundreds of raucous, screaming nesting gulls, white with gray wings. Manassie knew they would be here, and obviously was wholly familiar with every nook and cranny of the islands and the Baffin coast.

Freshwater streamed down from rocky headlands in a protected small cove, our destination for char. An occasional small plant tucked into a moist spot was the only visible plant life. A group of Inuit were there ahead of us, fishing with spinning rods, catching a lot of fish. Manassie explained that they were probably camping for several weeks somewhere up the Baffin coast, awaiting the arrival of the narwhals. Fred was crestfallen when he found that the only fly-fishing line he had was bereft of the forward shooting head. Casting rods would have to suffice, using daredevils again. The Inuit waved us to come over to them. After brief handshakes and instructions on how to fish here, they

FIGURE 44 Looking up into Fury and Hecla Strait.

departed in their motorized freighter canoe. Perhaps they had enough fish, or maybe they were just particularly mannerly, making way for us.

Char were everywhere, breaking water with their dorsal fins, feeding below the surface. In just two hours a large cooler was filled to the brim with beautiful fish. Some fish were thirty to thirty-five inches, and probably weighed ten to twelve or even fifteen pounds; the average must have been six or seven pounds. Manassie, Brad's wife, and the boy shared a pair of not very functional rods, and were happy to use our rods. Their absolute joy in hooking up with good fish was infectious. One plump char was cooked on the rocks and accompanied by the always available tea. The char was delicious, better than salmon, delicate, flavorful. The many char that we did not eat were taken to town by ATV at night, and distributed among Inuit families, delighting everyone.

The return trips from Baffin Island seemed longer, especially on the second trip when Manassie informed us the gas was low, and we would have to go very slowly to conserve fuel. It took many hours to get fifty miles back to camp, but that allowed time for peaceful contemplation and conversation. We saw only one other boat, a green freighter canoe with a single Inuk in it. If we ran out of gas, we inexplicably had no way to call or radio anyone, since the satellite phone was at a distant camp far down in Foxe Basin. Manassie allowed that once he ran out of gas and had to wait three days before someone picked him up. Even worse was another occasion when his motor failed, and he had to wait two weeks before he was found. This is big and quite empty country. We had only a day's supply of freshwater but plenty of char.

The seas were like glass, and we made barely any wake. The skies were a pure blue, and the border with the sea was indistinct, a continuum of blue across the visible world. The sun dropped lower, and gray clouds dominated the skies. At the edge of the water, beneath the cloud canopy, pink, orange, and yellow hues peeked through. Shades of silvery gray tinged by touches of orange reflected perfectly off of the mirrorlike waters, outlining the distant islands and shores up in Fury and Hecla Strait, another incredible display of crystalline-pure color and light.

The trips south into Foxe Basin evidently were more hazardous because Brad planned for a second boat to accompany us "for safety." Unfortunately, the engine on the second boat could not be started, so safety was put aside, and we traveled alone. A day's supply of food, some freshwater, a GPS device, and orange Arctic survival suits constituted our emergency gear. We headed for Manning Island, in search of the walrus herds. A bear might be there hunting walrus pups. Perhaps we could pick up the satellite phone, if marauding polar bears had not destroyed the camp. The weather was gray, with overhanging clouds through which Carolina blue skies or brilliant shafts of light appeared now and then. Ringed and bearded seals showed intermittently, and the bearded seals in particular seemed curious, following us and watching at a distance, reminding us of otters in the Rocky River below our home in North Carolina.

Brad's young son was eager to shoot a ringed seal (*natsiq*), and we spent much time in pursuit of one. Bullets splashed around the head of one poor animal that we hunted for an hour, but the seal was too clever, or injured, and we finally quit the chase. It seemed crazily reckless at times, as the boy perched on the bow, and Manassie gunned the boat when the seal was seen again. This water is cold, but the very agile boy kept his balance, and remained dry and enthusiastic.

Ice appeared on the horizon. The ice was reflected off the skies, a mirror image or "ice blink" seen more easily at a distance than the ice itself. The derelict radar tower at Hall Beach was just visible with good eyes, a forty-mile marker from Igloolik. Sadleq Island appeared and then disappeared to the east. Eventually, masses of floating pack ice materialized, and before long we were at the edge, seeking a way through. We tried for hours, without success. The massive ice was too densely packed, and we settled for cruising its boundaries in search of walrus pods.

Individual bits of ice were carved in every imaginable shape: seahorses, dinosaur heads, birds, strange creatures. Some ice was white, other pieces were tinged with blue. Dirty yellow-gray ice was a sign that walrus had rested there. Some of the floes were on top of each other, creating crazy angles and piles of ice. Within the ice, the waters were absolutely calm. Everything looked the same, and it was disorienting, with no sense of direction. Surprisingly strong tides pushed pieces of ice into each other. No land was visible; we were in the middle of a seemingly limitless ocean of ice, moving slowly.

Ready for lunch, we docked at a floating ice pack. Manassie marched to the top of an ice hill with his rifle, always watching for bear, and scouting for a way through the pack. Walking on the soft ice was easier than we imagined, but cold blue water showed through holes in the ice, and we were careful with our footing. Enjoying a potty break and lunch, we failed to notice the creeping ice, slowly moving closer, aiming to trap and crush our boat. Manassie's quick warning was sufficient to get us all on the boat and just barely reverse out of danger. We knew we escaped what could have been a major incident.

There was time for talk while we scouted the ice for walrus. Manassie was certain that the sea ice was thinner because of global warming. The ice used

FIGURE 45 A bearded seal lounging on the pan ice of Foxe Basin.

to be six feet thick, but now often was only two feet thick. Unlike Pauloosie, the elder whom Joyce met here in the depths of winter a year and a half ago, Manassie was completely comfortable talking about incipient change, and did not seem at all concerned. He enjoyed guiding, but was anxious to begin the narwhal hunt. If the narwhals came, he would leave to go after them. He had jobs in town at times, but his heart was out here. "I am an Inuk.... I like to hunt.... I went to Chesterfield with my girlfriend because of her job, but there were not many animals there, and they did not taste the same.... The caribou were small.... Even the seals tasted different."

We discussed myths. Some myths "are true, some are not... Elders taught us about them in school.... My girlfriend's stepfather was attacked by a sea monster, with long nose, webbed fingers, and covered in eider feathers. Her stepfather has a scar on his arm to prove it. Monsters like this live on the ice and under the sea, and can capture children.... Maybe they eat them." We liked Manassie's kind and patient manner, and appreciated his quiet competence, wiry strength, and self-assurance.

An occasional bearded seal (*ugjuk*) rested motionless on a bit of pan ice. They were wary. Turning the engine off and drifting slowly and silently down with the tide, we surprised a big ugjuk asleep on the ice. It looked up at us with large, deep, liquid eyes, its face framed with the large whiskers that give it its name. We could have touched it with a pole, before it slipped beneath the water. A beautiful, sleek, rich coat covered its massive body. It looked lonely and sad somehow, and entirely benign.

After searching for hours, we found hundreds of walrus (*aiviq*) close together. They have been here for millennia, providing a secure source of food for the Inuit. They were gathered in small and big clusters on tiny bits of ice that seemed much too small to support their weight. One pair sat dumbly without any intention of vacating their icy perch no matter how close we came, nor how long we stayed. Others remained warily on the ice with their pups perched protectively on top, huge bodies packed together. Fifteen or more animals were gathered in groups, only to slide as one into the water if we stayed too long or came too close. The pups appeared reluctant to leave the warmth of the massed family on the ice and had to be forced into the icy water. The walrus pup is dependent on its mother for about two years, while the seal pup is protected and fed for only a matter of weeks.

Their very small red eyes followed us closely, looking distinctly unfriendly. Some had large ulcers on their thick, wrinkled hides, apparently the result of turf battles, some of which we witnessed. Their two- to three-foot-long tusks helped to lift them up out of the water onto the ice, and were potent weapons when aroused. Some animals swam rapidly away in a huff, when our proximity forced them to vacate the ice. One big male swam straight for our boat as though to attack. Or maybe he thought our white boat was another icy perch. His intent made no difference; he was very large and very close, coming fast. Aggressive rogue males are well known, and can be a real threat to people in boats, or people or animals standing near the edge of an ice floe. Manassie jumped up in his orange survival suit, waving his arms. The walrus swam underneath, frightened perhaps by Manassie's apparition.

The walrus supported their massive bulk almost entirely on a diet of clams from the bottom of the seas. Floating slowly about on the ice took them over different waters, into new clam-rich areas. Loss of the ice by global warming would be a disaster, for they would no longer have the same access to rich clam beds. They were impressive in many ways, but ugly by conventional standards, and they were rank when approached downwind. Yet they could be beautiful, too. A permanent memory of the Arctic was the sight of walrus at a distance; peaceful brown blobs on white ice, floating on calm blue-gray seas under pastel skies.

"*Nanuq!*" Manassie yelled excitedly as he rammed the throttle forward, the engine roaring to life. We raced toward a distant mass of floating ice. What good eyes! We saw the bear only after we drew much closer. The bear was swimming, and turned to look at us. He dove, only to reappear inside the broken and rotting ice. Repeatedly he surfaced to glare at us, only to retreat within the ice island, trying to escape. Although we waited for an hour in hopes of seeing the whole of him, he never got up onto the ice. What we did see was impressive enough: a very big head and massive shoulders. We were very close, to his great discomfort. He seemed nervous, shaking his head, the water flying. He snorted, smacked his lips, and clicked his teeth—intimate sounds of an unhappy polar bear. Sometimes he dove and reemerged to peek at us like a little puppy, his head low on the ice. Walrus were visible behind him in the distance. He probably was living on the floating ice pack, hunting walrus and seals. In the water, he would be no match for an adult walrus, but a pup might be cut loose from the pack. Manassie was just as excited as we at the close encounter. The Inuit have great respect for nanuq: dangerous but worthy foes, tough, smart, persistent. Their many special adaptations to life in the Arctic are a wonder.[7] Finally, we broke it off, knowing it was a long way back to camp, and the seas were getting rougher under a growing wind. The big bear watched us for as long as we could see him.

FIGURE 46 A growling, teeth-clicking polar bear hiding amid rotting pack ice in Foxe Basin.

Traveling through these waters in a small open boat was exciting, and the only way to see the animals. There were no other boats within sight for the entirety of our time in the middle of Foxe Basin, seeking walrus and bear. We felt quite alone, and a bit vulnerable. The word *tariuraaluk* (out of sight of land) came to have personal meaning.

On the drive back to town from Brad's camp, along a narrow gravel road, there were a plethora of archaeological sites. Time was short, and we wanted to spend it with Jeena. After checking in at the Igloolik Inn, we visited Jeena at the quite modern and well-appointed health center. The nurses and other health workers spoke highly of her work ethic.

Jeena and Scott hosted us for a long evening at their home. Jeena said, "We have not had much time together…. We were so excited when you called…. Joyce and Fred are coming, not sure when but they are coming!" Slowly much was revealed. Although they readily answered our questions, we learned later that Inuit do not ask questions of elders, even in-laws. We Qallunaat were an exception, since we were not of their culture. We learned more of their courtship, their relationship, their families, their jobs, their hopes and frustrations. Jeena informed us she actually has sixteen names, bestowed on her by various members of the extended family, a means to tie them to each other, and to recall those gone to the far beyond. Passing names of the deceased on to the newly born enables their souls to find a new life, according to traditional belief.

Scott was raised in Hall Beach. Jeena allowed that Scott was a good father, loved their sons, and helped in the house. Scott was very supportive of Jeena's education, but her mother and father were not at all sure they wanted her to pursue further education in Iqaluit. First daughters are special. Moreover, Scott's dependability, leadership, and hunting skills were valued and needed by the extended family. Jeena was imprinted by a teacher in the third grade, and wanted to teach, but she also was being encouraged by nurses at the health center to become a nurse. All nurses were white and working on a rotation schedule, flying in for two or more weeks at a time. A full-time Inuk nurse would make a difference here. Pursuing a nursing degree would require a move to Iqaluit and another three years or more of school. A tough decision was put off for the moment.

Jeena was crocheting wool hats for each of us, and gave Joyce two pairs of earrings, one in the shape of an inuksuk, the other an ulu. Scott gave Fred his cotton ranger jacket and a very good knife. Fred attempted a response with the gift of his favorite Westport cap with a lobster pot on front, which Scott received with pleasure, but this seemed a mighty small gift compared to their generosity.

There were lots of people out on the streets at 10 p.m., including little kids on plastic skidoos. We met Annabelle, another of the women Joyce met here in January 2006, and stopped to talk. Annabelle was the female star of the movie *Snow Walker*, but was now unemployed. She hoped to sign on with the Mary River iron ore company. We ended the evening back at Jeena and

Scott's home watching *Snow Walker* on DVD. Annabelle was convincing in her role of a beautiful and heroic Inuk woman who was dying of consumption, who saved a rough white man after their plane crashed on the tundra. Cloaked in the warmth of friendship, we ambled to our hotel in the cold gray light of midnight.

On our last morning, we visited John MacDonald. John is a modern Knud Rasmussen, helping to preserve Inuit customs and skills. In response to our questions about Manassie's story of the eider-feathered creature, John opined this was how grandfathers convinced little children to stay off the ice. He was worried about increasing vandalism by some young people, which he attributed to alcohol and a recent problem with drugs in the community, including crystal meth. Dealers come intermittently with alcohol and drugs, leaving havoc in their wake. Another concern was an increasing split between the more affluent and the poor. Inuit with jobs could buy boats, skidoos, and ATVs. Because they worked, their boats and skidoos were used only on weekends. People without jobs felt this was unfair, and boats should be used by all. Talk drifted to other aspects of contemporary life in the High Arctic. John and Carolyn have lived there for many years, and clearly love the land and the people, but theirs is a real and not a romanticized world. Because this was the season to stock char for winter, he had to leave us to tend his char nets. Later we saw John greet his tiny brown Inuk friend Maurice with a huge smile, which Maurice returned in full. John's Inuktitut seemed entirely fluent, and the mutual love between them was obvious and moving.

Jeena appeared late for our final lunch at the Inn. She was withdrawn and appeared sad, quite unlike the happy person we left the night before. In a soft voice, she told us her father's best friend was beaten to death by his drunken and drug-crazed son the prior evening. Scott was in the boat to pick up her father at camp, to bring him back to town. We could not grasp the inhumanity of this tragedy. The body was at the health center. The two huge RCMPs who arrested the murderer were in the inn for lunch, sitting nearby. We wished we could help. The best we could do was listen. Jeena needed to talk, and she shared more of her own personal history, which seemed therapeutic somehow. Her mood brightened a bit gradually, but before very long we had to depart for our flight home. It seemed particularly sad to leave, and words failed. Jeena was a hope for the future, and would be a thoughtful and responsible teacher or nurse. She was a bridge from the past to the modern world, as comfortable with the Internet as hunting eggs on the tundra, or hunting seals on the ice. She was our friend, and we hoped for the best. It was not easy living in the eastern Canadian Arctic—torn between a distant, difficult, yet glorious past, and a present life with cultural contradictions and new challenges.

Over subsequent months and years, we corresponded frequently with Jeena by email, often several times a week. The relationship seemed important to her and certainly was important to us. Generally she was upbeat, and rarely talked about problems, nor complained. We lived vicariously through her messages, enjoying adventures of her family and community in the comfort of our home.

A man, his wife, and adopted son got lost on the frozen sea and were feared dead, but were located by a plane; their skidoo failed and they survived by building an igloo and using a seal-oil lamp for warmth, the old ways coming to the rescue. Scott and a friend brought back eight caribou after a five-day hunt, an eight-hour trip over frozen Foxe Basin by skidoo, although they were trapped on the ice for a day by a storm. Her father almost fell through a weak spot in the ice, but survived. Her father and brother went out on the April ice and saw hundreds of narwhals at the floe edge. Scott crashed his skidoo on the ice, resulting in hospitalization in Iqaluit with cracked ribs and internal injuries. To celebrate New Year's Eve with firecrackers and rockets, they traveled forty miles to Hall Beach over the frozen sea with their little children wrapped against the $-40°F$ cold in the skidoo. The fireworks were spectacular.

Pictures of her family and the town in winter ice and snow brightened our days in the heat of early summer in North Carolina. We sent pictures of dogwoods and azaleas in bloom, and of Easter egg hunts with the family. On occasion, we sent books, including Harry Potter books or DVDs, or other things that might be useful or that they wanted but could not get in Igloolik, including an artificial Christmas tree.

Jeena and Scott decided to get married. We finally were able to repay their generosity by sending formal black trousers, white shirt, and black bow tie for her husband-to-be on the occasion of their wedding. Such items were unavailable at the Northern store. Luckily, despite the inefficiencies of the mails in the Canadian Arctic, and the rush of Christmastime, the precious package arrived in time. She was grateful for our help, and sent photographs of the happy scene. Everyone looked grand.

Believing it was most important to stay near family, Jeena decided to concentrate on a teaching career. Teaching little kids would be a good way to help the community. She passed every course and graduated. Although a teaching position is not guaranteed, she was awarded a primary school teaching position. The position would help to solidify her family's financial situation, and would provide meaningful work. She was delighted with the good news.

We shared personal thoughts. Jeena wrote that she really never had similar white friends. She opined that she did not know about other people's homes, but here at her mother's home we could eat or drink anything without asking, because what was theirs was ours. And then, later in 2009, "Are you coming up to see the bowheads in June?"

Good question. How long can we keep going up there? But how can we not go?

12

FRIENDS OF THE FAMILY: A RETURN TO IGLOOLIK

I basked in the sun and drank in colours.... Every now and then an icy blast on my face reminded me that I was still on earth and had not drifted off into some idyllic nether world.

GEORGIA[1]

JULY 2010 For three years we waited to return to Nunavut. Our minds and hearts did not waver, but life has many facets, many responsibilities. Correspondence by email with Annie and Jeena remained steady, especially with Jeena, and occasional phone calls added spontaneity and spice. Although we have had less contact with our friends in Arviat, a phone call to Peter and Mary resulted in an invitation to visit.

"Why not?" said Peter cheerfully, and we agreed. We had not seen them since our fortieth anniversary trip seven years before. Their voices were warm, their enthusiasm for a visit infectious.

Annie emailed us that Joe was planning to enter his dog team in a race in the spring of 2010, from Rankin to Baker Lake, about two hundred miles. He has won many trophies in the past. We received her news as a possible visit opportunity, a chance to see the snows of April, and to renew our friendship. We could stop in Arviat along the way to see Peter and Mary. Fred was keen to experience winter in the North, but not knowing exactly when the race would take place presented a serious problem in planning. In September 2009 we tried hard to establish the date, contacting acquaintances in Rankin and Baker Lake. An Inuk lady on the phone at the Nunamiut Inn in Baker Lake was confused by our questions, but finally understood, and laughed heartily—at us rather than with us, we fear.

"They will decide when to race when they see what the weather is like! No, you can't make a reservation yet, it is much too early. We do not make reservations a year ahead of time." Once Qallunaat, always Qallunaat. We did not get to see the race.

Joe told us later that the race actually took place in May—he finished third—at about the time Fred was flying to a medical conference in Shanghai on a route that took him across Nunavut seventy miles south of Baker Lake. The constantly updated flight map of the route showed it went directly across the Kazan River. From thirty-five thousand feet, the clouds opened briefly, revealing black ice on Yathkyed Lake and the Kazan surrounded by an endless expanse of ridged and sculpted white. We paddled those waters in summer 2005, seeing caribou on their migration, wolves, musk-oxen, and many relics of former Inuit camps. No one else on the plane was at all interested in what lay below, apparently oblivious to the history and the grandeur of the land and its people. We had been there and knew it was still an untamed wilderness, one of few left in the world, a special place. From far above, it looked absolutely untouched. The newly-opened gold mine forty miles north of Baker Lake was not visible.

Annie emailed again just after Christmas, this time to tell us that she had been run over by a truck while she was walking to work in the dark and cold, rendering her black and blue from the waist down. Amazingly, no bones were broken, but she did sustain injuries. We worried about what medical care was

FIGURE 47 Jessie Oonark, *Untitled*, 1972, Baker Lake. Wall hanging symbolically depicting women (the ulu) and their role in regeneration (bird with her eggs). Men are a necessary but perhaps not a very large part of the whole (two male figures)! The bright colors were achieved by detailed stitching with embroidery floss over the cutout felt appliquéd images.

available to her, and we called. She did not seem angry, and was very thankful to be alive. We continue to marvel at how tough and resilient she is, always positive. By April she was much better and was back on the tundra, riding her skidoo out to their camp close to Baker Lake, hunting caribou and wolves. The hunt was successful, taking six caribou. Much of their diet depends on caribou, and they harvest about one a week over the course of a typical year.

The news from Igloolik was constant, with a plethora of shared photos. Jeena invited us to visit again. A trip in late June 2010 was aborted as a result of a decision of her extended family, including relatives from Hall Beach, to travel to Baffin Island to visit the grave of a deceased grandmother. Jeena later emailed that they saw the bowheads at about the time we planned to visit: "On our way to camp we were on the floe edge. I heard this loud noise coming from the water. I thought it was the ice breaking up, because it is that time of year. I looked downward and saw five bowhead whales. They were beautiful. On the way back we saw them again, but they were split up. It was beautiful."

Adaptability is a virtue. At the end of July 2010 we finally returned to Igloolik. Since we last visited, Jeena had a little daughter, and now had four children. She was well along in another pregnancy and Fred boned up on home delivery, fearful that his services might be needed. Thankfully they were not. Scott thought late July might be a better time to visit because travel over water by boat would be easier than travel over melting ice by skidoo. It took discipline and some imagination to pack for cold weather in the 102°F (39°C) heat of North Carolina.

Dense pan ice in Foxe Basin extended all the way to Hall Beach on the final leg of the two-day flight, but the floe edge had disappeared with the onset of summer. It was almost ice-free around Igloolik. Jeena and Scott did not appear at the airport until everyone else had departed, causing some anxiety on our part. One man was still there, loading a van with newly arrived supplies for the inn, and we hopped aboard. We should have been more patient, for on the way to town we saw a green Ford coming our way leaving a cloud of dust. Stopping, we happily embraced Jeena and Scott.

Their car battery failed, and they could not immediately find the jumper cables. The man from the inn seemed surprised that we were staying with Inuit in their home.

We spent two days in town while Scott finished assembling a new eighteen-foot aluminum boat. The home was modern and well-equipped, with one of the two bedrooms saved for us, but it was uncomfortably warm. What a pleasure to get outside into cooler air. Arising at 9 a.m., we had the streets to ourselves. The inn and the older hotel seemed empty. No other tourists were evident. Coffee was available at a small counter at the Co-op, where we were served by an emaciated-appearing older woman.

"The coffee for elders is free.... Take what you want." Elderly Inuit men at the next table offered to share their food with us, with big welcoming smiles. On the second morning the waitress also warmed up, asking whether we were there to make a documentary. When we told her that we were just visiting friends, she also seemed surprised. She offered that she spent ten years down south in Ottawa, where she lived on the streets much of the time. Now she was back home and happier, but needed to go south for a health matter soon. She did not look well.

The stay in town was an opportunity to visit acquaintances from our earlier trips. Our Inuk guide from three years ago, Manassie, was working in town, and seemed as delighted to see us as we were to see him. He seemed to chafe a bit at the confines of a regular job, but was full of joy. Annabelle, the starlet from *Snow Walker*, had left town for an unknown destination. Brad, the operator of the tour guide company that took us to the walrus herds, left town with his family after the hunters and trappers association ruled that he was disturbing the walrus too much. John and Carolyn MacDonald had retired and were living in Ottawa.

Two of Jeena's sisters drifted in and out, helping to care for her kids. Jeena gave us just-finished crocheted wool caps with our names in English and syllabic Inuktitut, each in perfect Carolina blue and white; kids called us by name on seeing our caps. A happy reunion at Jeena's mother and father's home was a chance to bestow gifts, including a fine old hunting knife from Fred's youth. We joined throngs of children at the town harbor in fishing for char and cod, and caught a large char, which we cooked.

After two days, Scott finished decking out the new boat, and we left for camp at 11 p.m., dressed in wool long underwear, wool caps, gloves, and parkas. Drifting pack ice was abundant, only two and a half weeks after the final breakup of sea ice. Clouds darkened the gray skies, and the sea was dark and flat. Despite the chill, there was much good cheer and talk during the hourlong trip. The camp slowly emerged on the stark and otherwise entirely empty stony shores; it had been enlarged since our visit three years before, with two new cabins. Scott departed, heading back to work for a day, leaving us with a loaded rifle but no boat. We were alone with Jeena and three of her kids. Our sleeping bags were laid out on skins covering a sleeping platform in the tiny cabin they call "room 26," formerly used by Jeena's mother for sewing skin clothing. Jeena occupied their newly finished cabin adjacent to ours.

A crash on our cabin door at 3 a.m. announced the intrusion of three Inuit hunters, who carried rifles and seemed amazed to find us in the little cabin. Joyce awoke first, sat bolt upright, and invited them in. There was commotion and muttering, while we all tried to figure out what was happening. Their engine failed, and they planned to sleep here until repairs could be made. After a few brusque words, they departed for the largest cabin. They had no way of knowing anyone was in camp, and were shocked at seeing elderly white people in the cabin. We joined them and tried to explain things, and soon we all were calmer.

The next day was peaceful after the hunters got their engine going and departed. By evening, however, activity multiplied with the arrival of Jeena's mother and father, her grandmother from Hall Beach, siblings including one of Jeena's brothers, two adoptive sisters, her brother's girlfriend and their two children—a total of sixteen Inuit across four generations. We were distributed in the three cabins and a sleeping and cooking tent, with a small tent for a community honey bucket. Miraculously, the honey bucket was

FIGURE 48 Family camp in golden evening light on Foxe Basin.

refreshed every morning. Sealskins and char were drying on racks; walrus and polar bear skulls were beside one of the cabins; scattered about were assorted skidoos and a Honda ATV which Jeena's mother bought after selling a polar bear skin that she was fortunate to be able to harvest two years ago.

"It was really big, and I was scared," Jeena's mother said.

Lots of char and fresh sealskins were kept refrigerated in boxes dug down into the permafrost. The camp was in excellent repair, and well equipped, including a generator that allowed electric lights in the cabins. Father kept a CB radio in the main cabin for use by stranded travelers from Igloolik or Hall Beach. Polar bears frequent the area, but none has caused trouble at the camp.

Much time was spent in discussion, and Mother in particular was a source of lore and stories. She informed us that her people are best described as Amitturmiut rather than Iglumiut, since the former term describes the various peoples who have long lived and hunted throughout the northern Foxe Basin area, and Iglumiut refers only to the people of Igloolik. There are many relatives in Hall Beach, drawn from surrounding camps fifty to sixty years ago by the construction of a big defense early warning ("DEW" line) radar site. One of Jeena's brothers was away at the moment, cleaning up the mess at a former DEW line radar site, so the Cold War still provided jobs for the Inuit.

"Do you know about the Queen of Igloolik? Her grave is just outside camp." We did know the story of Ataguttaaluk, who was forced to eat the frozen bodies of her dead children and husband to survive during a famine nearly one hundred years ago, She later became a respected camp leader. We found her grave not far away, stones piled on a gravel hilltop overlooking the sea. Terns dive-bombed us as we walked the beach, inspecting nearly six-foot-high chunks of sea ice left by the winds and tides.

"Would you like to go seal hunting?"

Of course! We dressed warmly, and departed in two boats at 4 p.m., expecting perhaps a four-hour hunt. It was 4 a.m. when we returned. Our journey took us through pack ice on a calm day. The broken sea ice often was colored a variety of greens and blues, and was dense in certain areas. The other boat saw a single bowhead whale, but we caught only a brief glimpse. Seals popped up intermittently, and two ringed seals were harvested, after many shots from the open boat. Clambering up onto floating ice and waiting for seals to appear was not fruitful. A visit to "tern island" was a welcome relief for several reasons, and everyone delighted in the abundance of eggs and young tern and eider duck hatchlings. Hordes of angry terns attacked from above until we left. Our hosts worried about hypothermia, but we were warm enough, barely. Braced by the cold air, we were not tired on return, and finally retired at 5 a.m.

The light was amazing every night, a kaleidoscope of colors from 10 p.m. to 3 a.m. As the sun slowly descended and glided along the horizon, everything on the ground was deep gold. Soon pure rose and pink exploded, glowing broadly across the wide-open skies and reflecting off the clouds onto the water. Scott's boat appeared to be moving through a wall of fire when it arrived from Hall Beach with Grandmother. Cabin windows were aflame with reflected color. The blue-gray skies behind the cabins were streaked with tendrils of soft rose. The combination of cold air and hot color made for an intense experience. We understood why the Inuit typically arise at noon in summer, going to bed about 4 a.m. At night there were fewer mosquitoes, and truly glorious colors. We adapted, and soon were awaking at noon also, welcomed by the fresh smell of bannock cooking.

Sunday was a day of rest. No hunting was allowed. At least eight young people were still sprawled under blankets and quilts on the sleeping platform in the big tent at 2 p.m., while Jeena's mother cooked char over a qulliq, and others ate, happily perched on boxes. There were no squabbles, and everyone had a role in camp life.

Meals were almost entirely country food, including caribou brought from their freezer in town, fresh seal meat from our hunt, and many meals of char. Our hosts are not much enamored of char, undoubtedly because they eat so much of it. While we were in camp, Jeena's brother netted sixty-two char that were either dried or saved to be frozen in town. Between us we added only four caught with casting rods; the char were stuffed with little shrimp and would not take the flies offered with a fly rod.

Seal tasted wonderful, a deep red-brown meat cooked in seal blood and vegetable soup, served with pickles and salt. We ate crowded around a plywood table perched on boxes on the gravel beach. Although Grandmother spoke no English, her eyes sparkled when Joyce deposited a tasty morsel on her plate. In thanks, Grandmother found some "macaroni" in the mix of seal parts on the serving plate, placing it on Joyce's plate. The boiled intestines were a tasty treat. Fred was the only person to eat the fat attached to the chunks of meat (everyone else carefully carved it off with an ulu), and he was the only one to suffer the explosive diarrhea that Mother cautioned could happen if one did not eat sweets afterward.

Our visit was over too quickly. As we departed camp at midnight under an intense red sky for the boat trip back to town, Jeena's father waved from shore and told us we were welcome back anytime. He is a strong man of few words, and his comment relieved us of apprehension about intruding into his family. Not willing to waste a moment, Scott and Jeena bypassed Igloolik and instead took us to some of their favorite haunts on the shores of Abverdjar, showing us secret coves and inlets where it is possible to escape from wind, and where seals are abundant. We imagined being Graham Rowley and coming upon this uninhabited island where he proceeded to dig many small ivory carvings of the Dorset culture. The water was a clear blue-green with large boulders perfectly outlined far below. Black rocky islands were framed in rose; some islands had pockets of soft green slopes that attract caribou. We gladly traded sleep and warmth for this rare experience with our Inuit friends.

At the airport, a handsome young man in a First Air uniform called out to us, "Hi, Fred and Joyce." Jeena explained that he was one of the hunters whom we met at 3 a.m. so abruptly and rudely at camp. Trying to collect ourselves, we hoped our manners were as gracious as his.

Jeena and Scott each said, "Thank you for coming." It was we who needed to thank them for allowing us entry into their personal lives.

After we left, Jeena reported that they had very successful walrus and caribou hunts. She was thrilled that one of her sons shot his first caribou. As is the normal practice here, she flew hundreds of miles to Iqaluit to await birth of her baby. Scott was able to accompany her, and they enjoyed a three-week holiday, alone without kids for the first time in their marriage. Sadly, Scott had to fly home just before the baby was born to comfort the wife of his best friend, who committed suicide. Jeena told us about this by email. Shortly afterward, she sent pictures of her beautiful new baby girl, and she was ecstatic. The pictures showed two of her sons holding their new sister, beaming with delight. The juxtaposition of new life with an unnecessary death, of joy and despair, was poignant.

13

REFLECTIONS

*There is the misconception that we have reached our destination
the moment we grow old, but... we are still traveling
toward those destinations, still beyond our reach.*

TAN TWANG ENG[1]

FEBRUARY 2011 To visit with Jeena and Scott, we must fly from North Carolina to Ottawa, to Iqualuit, and finally to Igloolik, always a two-day journey—that is, if there is no fog blanketing the island. If we want to visit Annie and Joe, we must fly from our home, through Minneapolis, staying overnight in Winnipeg before flying through Churchill, Arviat, and Rankin Inlet to reach Baker Lake. It is similarly difficult to get to Pangnirtung. Traveling in the North is inconvenient, difficult, and expensive. Why do we yearn to go to the North? Is there not beauty everywhere?

Nunavut is different from any other place we have been.

The people are particularly welcoming. We are pleased when we walk the few streets of these hamlets and are greeted happily by total strangers, or when we ask to meet a well-known Inuit artist and are invited into her home and served tea, bannock, or *maktaak* (white whale skin and blubber—a delicacy). We are encouraged by Inuit perseverance and courage as they strive toward a meaningful future: adapting to new language and customs, establishing and achieving new goals. The Inuit are resolute, patient, strong, and capable; they are survivors. Knowing the Inuit connects us in some manner to our own dim past, when our immediate ancestors lived on the land as farmers, and earlier as hunter-gatherers. The Inuit help us to understand our own origins.

Living on the land fulfills a spiritual need. We experience one of the last great wild areas on earth. The rivers run free and clean; the lakes are wide and wondrous with wind; the land is pristine, not yet overwhelmed by civilization and its trappings. On long walks on the open tundra, or paddling in a canoe, we are alone with our thoughts. Trivial problems disappear. The air is clear and bright; sunsets and sunrises seem timeless as the sun rotates slowly along the wide-open horizon. Our fellow travelers are handsome animals, some potentially dangerous, others awesome in their numbers—all magnificent. There is a sort of magic in the air.

The beauty of the land, animals, and people bring us new appreciation of Inuit art.

The carvings, prints, and wall hangings enable us to experience yet again the peace and harmony that envelop us in the North. Ruth Qaulluaryuk's two birds hover over our bed protecting us; Ruth's abstracted bird emerging from tundra stimulates us to observe more carefully. Eva Ikinilik Nagyougalik's wall hanging of tundra colors and caribou tracks takes us back to the freedom and beauty of the Barren Grounds. Janet Kigusiak's *Char Hanging to Dry* recalls our personal time with Jeena and her family at their Foxe Basin Camp. Besides appreciating their beauty, we hope our children and grandchildren will feel the essence of a special people and place in these creative expressions.

For us, the answer as to why the North is such a draw also relates to the intimacy that comes from sharing. Each of our works of art was chosen by us as a team, and the trips were planned and executed as a team. In the canoes, on the land, and in the galleries, we found new satisfactions as a couple.

Togetherness gave these adventures real emotional power. Awareness of the limits of time, something that comes more with age, added to our appreciation in ways that we might not have understood in our youth.

Although we feel connected to a world that at first glance is immutable, the North is not free of serious problems. As for people of the South, conflicts and problems exist in Inuit society: drugs and alcohol, loss of traditional roles, unemployment, inadequate education, depression and social disruption, insufficient housing. What concerns us most, however, is that the environment might be changing in ways that are hard to reverse. Many caribou herds are in dramatic decline, including most of the migratory herds of the open tundra west of Hudson Bay: the Beverly, Qamanirjuaq, Ahiak, Lorillard, Wager Bay, Bathurst, Cape Bathurst, and East and West Bluenose herds.[2] The Beverly herd we saw on the Thelon is down by 98 percent, and the Bathurst herd by almost as much, according to excellent data from 2009.[2] Other herds are somewhat less affected, explaining why Annie, Joe, Scott, and others have yet to notice a serious decline in their hunting success. Natural expansions and contractions of the caribou herds have a thirty- to forty-year periodicity. The last peak was in about 1970, so this decline may be part of a normal cycle. Recovery cycles, however, may be threatened by warming weather,[3] and by excessive hunting and mining. The number of leases and permits for mining diamonds, uranium, and gold in the region of the Bathurst and Beverly herds' calving grounds is rapidly increasing.[2] Drilling for oil and gas buried deep in the Arctic carries risks of spills, pollution, and habitat damage. A new road is being proposed to connect Baker Lake, Rankin Inlet, and Arviat, to the south, for the purpose of fostering land development. Although it undoubtedly will promote job creation, it will further threaten the ecosystem. Nunavut offers one of the last chances we humans have to preserve a piece of the world as it was created. If the last wild places are perturbed, it will be hard for all of us to know our origins, to appreciate what it is we seek to preserve and why, to understand our need to have respite, and to feel safe.

The lure of the North is powerful. Soon we will be paddling down the lower Back River through the former camps of the Utkuhiksalingmiut; of Luke Anguhadluq, Jessie Oonark, Tuu'luq, and our friends Ruth and Josiah Nuilaalik; a place of big char and big dreams. What better way to celebrate our 48th anniversary than to paddle across the Arctic Circle, bathed in the rich colors of the late Arctic summer? One more time, before it is too late.

FIGURE 49 *Off Again Canoeing*. Jessie Oonark's 1967 drawing of crossing a river in skin rafts. The caption, written in syllabics, cautions, "Be careful...Do not tip over and drown." Baker Lake.

ACKNOWLEDGMENTS

We are thankful for the many courtesies and knowledge shared with us by our Inuit friends and acquaintances. Many people let us into their lives and generously shared their thoughts as well as the hospitality of their homes and food. We are indebted to Hanna and her mother, Ruth Qaulluaryuk, and her late father, Josiah Nuilaalik, of Baker Lake, who shared laughs and stories with us. Sadly, Josiah is now gone, as are many other elders whom we met. Annie and Joe were extraordinarily kind to us during our week in their Baker Lake home and in their tundra camp. Annie shared many personal stories, both of the old days and of current life; at her request, we changed their names to protect their privacy. We omitted last names of all the Inuit whom we met, except for those who are already in the public domain as well-known artists. Jeena and Scott allowed us to enter their family in an intimate way, and to spend many wonderful days with them in town and in camp. Jeena and Annie were particularly generous in sharing their lives via the Internet. Mary and her family in Arviat were hospitable beyond a reasonable expectation; their immediate and unqualified friendship made our visit to Arviat memorable. The Maniapik family in Pangnirtung were extremely welcoming, particularly Lucy. Many other Inuit not named were kind to talk with us in the various hamlets of Nunavut, and added immensely to our experiences. To all of these people and to each of the many artists whom we met in their homes in Arviat and Baker Lake, we say *qujannamiik* (thank you).

Qallunaat art gallery owners and dealers taught and guided us at certain points in our journey. They promoted and developed markets for the art of the Inuit, and often provided assistance to individual artists. Many gallery people became friends, including Judith Varney Burch, Pat Feheley, John Houston, Judy Kardosh, Mark London, Derek Norton, Nigel Reading, and the late Faye Settler. Fellow collectors made our lives richer in many ways. Scholars of Inuit art added to our learning and enjoyment, particularly Darlene Wight of the Winnipeg Art Gallery. Marie Bouchard generously shared her knowledge of wall hanging art and artists, adding immensely to our appreciation of art and life in Baker Lake.

Many others share our love for the North. Our guides for the paddles into the Barren Grounds, Alex Hall and Rob Currie, added so much to our experiences that a simple thank-you seems totally inadequate. Both Alex and Rob led by example in a manner that the Inuit would admire. Our fellow paddlers also added greatly to our experiences, and in many ways were indispensable. The late Stu Mackinnon was an inspiration; bless his memory. Dale Vermillion's stories of his three canoe trips on the Barren Grounds first aroused our interest in similar adventures, and ultimately in our graduation from cruise ships to canoes. Don Sessions and his son, Lee, loved sharing stories of their extensive paddles on the Barren Grounds. John and Carolyn MacDonald generously shared their hospitality and their wisdom. Dorothy Harley Eber was more than kind to guide us in preparing this manuscript. Without the help of H. G. Jones, much of this tale never would have happened. He also read an early draft of this work, as did fellow paddler Rand, Sue Grieff, Betsey Maker, and James Anderson.

The authors took all photographs unless otherwise noted. The works of Inuit art are from the authors' personal collection. The Baker Lake prints are reproduced with permission of Inuit Art Services, Inuit Art Foundation. Maps were drawn by Craig Dalton, doctoral candidate in geography at the University of North Carolina.

NOTES

CHAPTER 1

1. Swinton, *Sculpture of the Inuit*, 244.
2. Many authors have discussed the quality of Inuit art. See Hessel, *Inuit Art*; Millard, "On Quality in Art": Swinton, *Sculpture of the Inuit*. Hessel and Swinton are perhaps the two best introductions to the art of the Inuit. For a more general discussion of the origins of art and its definition, see Dutton, *Art Instinct*.
3. For a general review of the art in Arviat, see Hessel, *Inuit Art*, and Swinton, *Sculpture of the Inuit*. Availability of different types of carving stone and its impact on regional styles of sculpture is discussed in Gustavison, *Northern Rock*. For original articles relating to particular artists in Arviat, see Hessel, "Arviat Stone Sculpture," and Kunnuk and McGrath, "Lucy Tasseor Tutsuitok."
4. Baker Lake art is discussed admirably in Swinton, *Sculpture of the Inuit*, among many others.
5. Discussions of the art of sewing on cloth (wall hangings, or works on cloth) appear in Blodgett and Bouchard, "Jessie Oonark"; Bouchard, *Marion Tuu'luq*; and Fernstrom, *Northern Lights*.
6. James Houston's life among the Inuit and his role in stimulating and promoting Inuit art is described in Houston, *Confessions of an Igloo Dweller*.
7. For histories and analysis of the origins of the Inuit in Canada, see Fossett, *In Order to Live Untroubled*, and McGhee, *Last Imaginary Place*.
8. Mowat, *People of the Deer*.
9. Harrington, *Face of the Arctic*.
10. Steenhoven, "Ennadai Lake People 1955."
11. McGhee, *Last Imaginary Place*.

12. Fossett, *In Order to Live Untroubled*.
13. Relocations of Inuit to the high Arctic are discussed extensively in McGrath, *Long Exile*, and Tester and Kulchyski, *Tammarniit*.
14. Spalding, with Kusugaq, *Inuktitut*. Inuktitut spelling is variable, due to differences in local dialects and difficulties in translating the sometimes guttural sounds into English. We generally adopted the spelling in this dictionary, which is based principally on the dialect of the region on the west coast of Hudson Bay. An exception is Qallunaaq, an alternate spelling to the favored Qablunaaq of Spalding. Some English authors use Kabloona instead. All mean "white man" or "European." Plural of Qallunaaq is Qallunaat (white men).
15. For a discussion of the creation of Nunavut, see Michell and Tobin, "Newest Member of Federation."
16. Global warming is much discussed. The Arctic summer ice cap is shrinking. Disappearance of the ice would have serious effects, including loss of the reflective properties of ice that help to cool the earth in summer. Animals would suffer, and some might disappear. For more thorough discussion, see Kerr, "Arctic Summer Sea Ice Could Vanish Soon but Not Suddenly"; Screen and Simmonds, "Central Role of Diminishing Sea Ice"; and Post et al., "Ecological Dynamics across the Arctic."

CHAPTER 2

1. Marsh, *People of the Willow*, 11.
2. Many excellent books recount the explorations of the Canadian North, including Kane, *Arctic Explorations*; Green, *Peary*; Jenness, *People of the Twilight*; Nanton, *Arctic Breakthrough*; Palsson, *Writing on Ice*. Others are cited elsewhere in this volume.
3. Parry, *Journal of a Second Voyage*. Holding this large volume—written shortly after the expedition to Foxe Basin and what is now termed Fury and Hecla Strait, with its detailed etchings of the Inuit and careful descriptions of the ice, animals, and weather—is to come as close as possible to firsthand experience. Parry hugely enjoyed the Inuit of Igloolik, who were very welcoming.
4. Brief summaries of the voyages of the Norse to Greenland and Newfoundland are found in Haugen, *Voyages to Vinland*.
5. Kent, *N by E*, 165. Wood-block prints of his aborted sailing trip ending in Greenland are many, and his writing is powerful.
6. Mirsky, *To the Arctic!*
7. Rasmussen, *Across Arctic America*. Rasmussen and his small Danish team conducted a marvelous anthropological, geological, and scientific exploration of Arctic Canada and Alaska, published in multiple volumes but summarized here. This is an essential starting point for students of the Inuit.

8. The voyages of Henry Hudson, including the final voyage into what is now known as Hudson Bay, the mutiny of his disgruntled men, the undoubted deaths of Hudson and others who were put adrift in a small boat, and the subsequent deaths of many of the principle mutineers at the hands of the Inuit at Digges Island are described well in Mancall, *Fatal Journey*. We saw the same masses of thick-billed murres at Digges Island as Hudson and his men off of Northwest Arctic Quebec. Mancall's also is a marvelous recounting of the repeated attempts to find a northwest passage in the early seventeenth century.
9. Savours, *Search for the North West Passage*.
10. Freuchen, *Arctic Adventure*, 349–54.
11. Gustavison, *Northern Rock*, 72–73.
12. For a discussion of Peter Pitseolak and his photography, see Eber, "Peter Pitseolak and the Photographic Template," 53–59.
13. McGhee, *Last Imaginary Place*.
14. Mowat, *People of the Deer, Desperate People, Walking on the Land*.
15. See Driscoll, *Inuit Amautik*, 11. "Amautik" refers to a woman's parka with a pouch or full hood for carrying a baby. An inner parka, or *atigi*, had soft fur in contact with the wearer's body and was often heavily beaded on the outside. An outer parka without carrying hood, the *qulittaq*, was worn by unmarried women.

CHAPTER 3

1. Rasmussen, *Across Arctic America*, 157.
2. McGhee, *Last Imaginary Place*.
3. Verne, *Adventures of Captain Hatteras*.
4. Among many Arctic adventures, the following are particularly poignant: Edinger, *Fury Beach* (as described in this book, Captain John Ross and his men observed the amazing skill of Inuit in geography and mapmaking); C. Stuart Houston, *To the Arctic by Canoe*; McGoogan, *Fatal Passage*. McGoogan tells the story of John Rae, who believed in traveling light, living as the Inuit, and who covered long distances on foot. He became infamous among the English because he uncovered the unwelcome first evidence that Franklin's men resorted to cannibalism.
5. McGhee, *Last Imaginary Place*, 111.
6. Tester and Kulchyski, *Tammarniit*.
7. McGrath, *Long Exile*.
8. For a general history, see Savours, *Search for the North West Passage*. For insight into the prolonged search for the lost Franklin expedition, see Kane, *Arctic Explorations*.
9. Eber, *Encounters on the Passage*. Interviews with elders uncover the oral history of the early encounters.

10. Beattie and Geiger, *Frozen in Time*.
11. Mirages and other optical illusions are common in the bent light of the Arctic. See Pielou, *Naturalist's Guide to the Arctic*.
12. Freuchen, *Arctic Adventure*, 349–54.
13. Simon, *North into the Night*, 98, 100.
14. Scherman, *Spring on an Arctic Island*.
15. Blais et al., "Arctic Seabirds Transport Marine-Derived Contaminants."
16. Kelly et al., "Food Web-Specific Biomagnification of Persistent Organic Pollutants."
17. Cone, *Silent Snow*.
18. See, in particular, Ehrlich, *This Cold Heaven*.
19. The richness of the waters in Lancaster Sound and Eclipse Sound, and the biology and behaviors of the narwhals that live there, are well described in Lopez, *Arctic Dreams*.
20. We inverted the order of the occurrence of the second cruise and the visit to Pangnirtung, presenting the visit to Pangnirtung (chap. 4) after the cruise through the High Arctic (chap. 3). The visit to Pangnirtung actually preceded the High Arctic by two months; it made more sense for the story to place the cruises ahead of visits to the hamlets.

CHAPTER 4

1. Roe, "I Know My Life Was Happier, Freer."
2. Hallendy, *Inuksuit*. This is the best—and most beautiful—book about inuksuit.
3. Eber, *When the Whalers Were Up North*.
4. Boas, *Central Eskimo*.
5. MacDonald, *Arctic Sky*.
6. Hankins, *Sunrise over Pangnirtung*.
7. Pryde, *Nunanga*, 82–83. An account of raucous days in the wild North.
8. Von Finkenstein, *Nuvisavik*. An exhibition catalog and discussion of weaving in Pangnirtung.
9. Grenfell, *Adrift on an Ice-Pan*. Grenfell was a physician who committed most of his professional life to serving the poor fishermen along the Labrador and Newfoundland coasts. This episode endeared him to the public and helped in fund raising for his mission.
10. Loomis, *Weird and Tragic Stories*.

CHAPTER 5

1. Pryde, *Nunanga*, 279.
2. Bouchard, *Marion Tuu'luq*.
3. Rasmussen, *Across Arctic America*, 70.

4. Kerr, "Geologists Find Vestige of Early Earth—Maybe World's Oldest Rock."
5. MacDonald, *Arctic Sky*.
6. Mowat, *Desperate People*.
7. Hessel, "Arviat Stone Sculpture."
8. For discussions of Inuit songs and poetry, see Rasmussen, *Across Arctic America*, and Lowenstein, *Eskimo Poems from Canada and Greenland*.
9. Lowenstein, *Eskimo Poems from Canada and Greenland*, 107.
10. Rasmussen, *Across Arctic America*, 34.
11. Pangnark's art is discussed in Hessel, *Inuit Art*, and Swinton, *Sculpture of the Inuit*.
12. Qamukaq, "Martina Anoee."
13. For a discussion of the ancient art of throat singing, see Paartridge, "Throatsinging."
14. Anguhadluq's drawings are discussed fully in Cook, *From the Centre*. His long walk into town off the tundra with his daughter Ruth was captured on film in Fransen, *Memories from Nunavut*, 60.
15. Blodgett and Bouchard, "Jessie Oonark."
16. Rasmussen, *Across Arctic America*, 199.
17. Briggs, *Never in Anger*.
18. Mannik, *Inuit Nunamiut*.
19. Fernstrom, *Northern Lights*.
20. Von Finkenstein, "Miriam Qiyuk."
21. Nasby, *Irene Avaalaaqiaq*.
22. Burt, *Barrenland Beauties*. Excellent photos, and a limited translation of names into Inuktitut.

CHAPTER 6

1. Rowlands, *Philosopher and the Wolf*, 109.
2. Hall, *Discovering Eden*. Funny, terse, full of essential lore for would-be paddlers in the Arctic, and imbued with his love of the Barren Grounds.
3. For a discussion of the physics and odd effects of light in the Arctic, including various mirages and difficulties in estimating distance and size, see Lopez's wonderful book *Arctic Dreams*.
4. Musk-ox behaviors are discussed in ibid.
5. Seton, *Arctic Prairies*.
6. Powell-Williams, *Cold Burial*.
7. Mackinnon, "Garry Lake."

CHAPTER 7

1. Pelly and Hanks, *Kazan Journey into an Emerging Land*, ix.
2. Lentz, "Inuit Ku."
3. Arctic Cairn Notes, *Canoeist's Reflections on the Hanbury-Thelon and Kazan Rivers*.
4. Grinnell, *Death in the Barrens*.
5. Details on former Inuit sites on 30 Mile Lake and the lower Kazan are found in Stewart et al., "Archeology and Oral History of Inuit Land Use on the Kazan River." In figure 16, Irene Tiktaalaaq Avaalaaqiaq's family's camp is located at Piqqiq.
6. Pelly organized and led a group of young people from around the globe who paddled down the entirety of the Kazan. Their notes and observations are an inspiration, and suggested to us that we could do the trip, too. Of course, they were young and we were not. See Pelly and Hanks, *Kazan Journey into an Emerging Land*.
7. Rasmussen, *Across Arctic America*, 61.
8. Ibid., 66.
9. Hall, *Discovering Eden*.
10. The sculptor George Tataniq expounded on his former life on the lower Kazan and Forde Lake, including hunting caribou from kayaks, in Mannik, *Inuit Nunamiut*, 225–26.
11. Webster, *Harvaqtuurmiut Heritage*, 26.
12. The uses of inuksuit, and what they meant to the Inuit, are discussed by George Tataniq in Mannik, *Inuit Nunamiut*, 231–32.
13. Klevens et al., "Invasive Methicillin-Resistant *Staphylococcus aureus* Infections in the United States." In 2005, ninety-four thousand people were hospitalized with such infections in the United States, with eighteen thousand deaths. Only 7 percent of skin infections resulted in such complications, thankfully for us.
14. Webster, *Harvaqtuurmiut Heritage*, 31.
15. Ibid., 32–33.
16. Ibid., 11.
17. Currie, *Wilderness Spirit Newsletter*.
18. Curtis, *Cave Painters*.
19. Shepard, *Coming Home to the Pleistocene*.

CHAPTER 8

1. Houston, *Confessions of an Igloo Dweller*, 116.
2. Back noted many Inuit camps along the river termed Thlew-ee-choh (big fish river), now named for him following his epic trip down and back up the entirety of the river

in 1834. Inuit camps were noted for the first time near the confluence of the "wild river" and the Back River, and became more numerous downstream. Near Chantrey Inlet, he encountered Inuit notable for their tatooed women, cleanliness, and honesty.
3. Houston, *James Houston's Treasury of Inuit Legends*, 205.

CHAPTER 9

1. McLennan, unpublished poem.
2. Rasmussen, *Across Arctic America*.
3. Parry, *Journal of a Second Voyage*.
4. Robinson, *Atanarjuat*.
5. MacDonald, *Arctic Sky*.
6. Pielou, *Naturalist's Guide to the Arctic*. A useful, short book with concise discussions of some of the weird visual phenomena in the Arctic, and reasons for their occurrence.
7. Gunderson and Kunu, "Giving Inuit a Voice."
8. Rowley, *Cold Comfort*.
9. Clewes, *Thule Explorer*, 100.
10. Simon, *North into the Night*, 98.
11. Rasmussen, *Across Arctic America*, 137.
12. Rowley, *Cold Comfort*, 69.
13. Fox, "Women Helping Each Other."
14. Among many writings about art in Igloolik, see Gunderson, "Luke Airut," and Wight, "Germaine Arnaktauyok."
15. Jenness, "Eskimo Art."
16. McDougall, "Jenness on Eskimo Art."
17. Billson and Mancini, *Inuit Women*.

CHAPTER 10

1. Bennett and Rowley, *Uqalurait*, 357–59.
2. Ibid., 351, 353.
3. Arctic wildflowers and their uses are discussed in Burt, *Barrenland Beauties*, and Pielou, *Naturalist's Guide to the Arctic*.
4. Rasmussen, *Across Arctic America*, 63. Also see Bennett and Rowley, *Uqalurait*, 342–59, for a discussion of the culture of the Ahiarmiut.
5. Bennett and Rowley, *Uqalurait*, 342.

CHAPTER 11

1. Ehrlich, *Future of Ice*, 104.
2. Rowley, *Cold Comfort*.
3. Freuchen, *Arctic Adventure*.
4. McGhee, *Last Imaginary Place*.
5. Lopez, *Arctic Dreams*, offers an excellent description of some of the Dorset artifacts discovered near Igloolik.
6. Parry, *Journal of a Second Voyage*.
7. The behaviors and special adaptations of the polar bear, a recent evolutionary derivative of brown bears including grizzlies, are discussed sympathetically in Lopez, *Arctic Dreams*.

CHAPTER 12

1. Georgia, *Georgia*, 83.

CHAPTER 13

1. Eng, *Gift of Rain*, 18.
2. The Beverly herd, which clusters around the Thelon watershed, has declined from an estimated size of 276,000 animals in the 1990s to a low of about 5,000 currently. The evidence is based on repeated aerial counts. As few as 15 percent of cows are pregnant or with calf, far lower than the usual 85 percent. The Bathurst herd north and west of Baker Lake once included 470,000 animals, but has been reduced to about 30,000 currently (a 94 percent decline). The pregnancy rates are very low there, too. The smaller Cape Bathurst herd has dropped from about 20,000 to 2,000 animals. The more westerly Bluenose-East and Bluenose-West herds also are declining, from about 112,000 to 20,000, and 120,000 to 67,000 animals, respectively. Numbers of caribou are taken from several sources. A rich resource is Hummel and Ray, *Caribou and the North*. Data on many herds are found in Adamczewski et al., "Decline in the Bathurst Caribou Herd, 2006–2009." Additional data and news about the herds are found in the Beverly and Qamanirjuaq Caribou Management Board Twenty-seventh Annual Report. The BQCMB reports are issued roughly twice a year and contain links to many other stories and news. These documents also are rich sources of information about mining and leases in the calving and postcalving grounds of the Beverly, Bathurst, and Qamanirjuaq herds. A general survey of the world's populations of reindeer and caribou is found in

Vors and Boyce, "Global Declines of Caribou and Reindeer." For a particular discussion of the vicissitudes of the Qamanirjuaq herd, see a report by Wakelyn, "Qamanirjuaq Caribou Herd."

3. There are several mechanisms by which warming could contribute to caribou declines. Earlier emergence of the most nutritious foods on their calving grounds may mean that their migration arrives too late. Thawing and freezing in winter affect their ability to get to winter lichens through the snow; their hoofs and shovel antlers are not equipped to penetrate ice. Longer and hotter summers may increase numbers of their insect predators, including mosquitoes, black flies, nose bot flies, and the warble flies that bury eggs under their skin. As caribou decline, so do tundra wolves that depend on caribou for their survival.

BIBLIOGRAPHY

Adamczewski J., et al. "Decline in the Bathurst Caribou Herd, 2006–2009: A Technical Evaluation of Field Data and Modeling." Draft Technical Report. Environment and Natural Resources, Government of Northwest Territories, Yellowknife, NWT, December 2009.

Anderson, Alun. *After the Ice: Life, Death and Geopolitics in the New Arctic.* New York: HarperCollins, 2009.

Arctic Cairn Notes. *Canoeist's Reflections on the Hanbury-Thelon and Kazan Rivers.* Toronto: Betelgeuse Books, 1997.

Back, George. Narrative of the Arctic Land Expedition to the mouth of the Great Fish River, and along the shores of the Arctic Ocean in the years 1833, 1834, and 1835. London: John Murray, 1836.

Beattie, Owen, and John Geiger. *Frozen in Time: Unlocking the Secrets of the Franklin Expedition.* New York: E. P. Dutton, 1987.

Bennett, John, and Susan Rowley, eds. *Uqalurait: An Oral History of Nunavut.* Montreal: McGill-Queen's University Press, 2004.

Beverly and Qamanirjuaq Caribou Management Board. Twenty-seventh Annual Report 2008–2009. www.arctic-caribou.com/PDF/BQCMB_2008_2009_Annual_Report.

Billson, Janet, and Kyra Mancini. *Inuit Women: Their Powerful Spirit in a Century of Change.* Toronto: Rowman & Littlefield. 2007.

Blais, Jules M., et al. "Arctic Seabirds Transport Marine-Derived Contaminants." *Science* 309 (2005): 445.

Blodgett, Jean, and Marie Bouchard. "Jessie Oonark: A Retrospective." Winnipeg Art Gallery, 1986.

Boas, Franz. *The Central Eskimo.* Lincoln: University of Nebraska Press, 1964. (Reprint of the original article published in the Sixth Annual Report of the Bureau of Ethnology. Washington DC: Smithsonian Institution, 1888.)

Bouchard, Marie. *Marion Tuu'luq.* Ottawa: National Gallery of Canada, 2002.

Briggs, Jean L. *Never in Anger: Portrait of an Eskimo Family.* Cambridge, MA: Harvard University Press, 1970.
Burt, Page. *Barrenland Beauties: Showy Plants of the Canadian Arctic.* Yellowknife, NWT: Outcrop Ltd., 2000.
Clewes, Rosemary. *Thule Explorer: Kayaking North of 77 Degrees.* United States: Hidden Brook Press, 2008.
Cone, Marla. *Silent Snow: The Slow Poisoning of the Arctic.* New York: Grove Press, 2005.
Cook, Cynthia Wayne. *From the Centre: The Drawings of Luke Anguhadluq.* Toronto: Art Gallery of Ontario, 1993.
Currie, Rob. *Wilderness Spirit Newsletter*, November 29, 2005.
Curtis, Gregory. *The Cave Painters: Probing the Mysteries of the World's First Artists.* New York: Anchor Books, 2007.
Driscoll, Bernadette. *The Inuit Amautik: I Like My Hood to Be Full.* Winnipeg: Winnipeg Art Gallery, 1980.
Dutton, Denis. *The Art Instinct: Beauty, Pleasure and Human Evolution.* New York, Bloomsbury Press, 2009.
Eber, Dorothy. *Encounters on the Passage: Inuit Meet the Explorers.* Toronto: University of Toronto Press, 2008.
———. "Peter Pitseolak and the Photographic Template." In *Imaging the Arctic.* Edited by J.C.H. King and Henrietta Lidchi. London: British Museum Press, 1998.
———. *When the Whalers Were Up North. Inuit Memories from the Eastern Arctic.* Montreal: McGill-Queens University Press, 1989.
Edinger, Ray. *Fury Beach: The Four-Year Odyssey of Captain John Ross and the Victory.* New York: Berkley Books, 2003.
Ehrlich, Gretel. *The Future of Ice: A Journey into Cold.* New York: Pantheon Books, 2004.
———. *This Cold Heaven: Seven Seasons in Greenland.* New York: Pantheon Books, 2001.
Eng, Tan Twan. *The Gift of Rain.* New York: Weinstein Books, 2008.
Fernstrom, Katharine. *Northern Lights: Inuit Textile Art from the Canadian Arctic.* Baltimore Museum of Art Catalog, 1993.
Fossett, Renee. *In Order to Live Untroubled: Inuit of the Central Arctic, 1550 to 1940.* Winnipeg: University of Manitoba Press, 2001.
Fox, Matthew. "Women Helping Each Other." *Inuit Art Quarterly* 13, no. 1 (1998): 13.
Fransen, Bernard. *Memories from Nunavut.* Penticton, BC: n.p., 2002.
Freuchen, Peter. *Arctic Adventure: My Life in the Frozen North.* 1935; repr. Guilford, CT: Lyons Press, 2002.
Georgia. *Georgia, an Arctic Diary.* Edmonton: Hurtig Publishers, 1982.
Green, Fitzhugh. *Peary: The Man Who Refused to Fail.* New York: G.P. Putnam's Sons, 1926.
Grenfell, Wilfred T. *Adrift on an Ice-Pan.* Boston: Houghton Mifflin, 1909.

Grinnell, George. *A Death in the Barrens*. N. Ferrisburg, VT: Heron Dance Press and Art Studio, 2006.

Gunderson, Sonia. "Luke Airut: Igloolik's Carving Wizard." *Inuit Art Quarterly* 21 no. 3 (2006): 11–15.

———, and Zacharias Kunu. "Giving Inuit a Voice." *Inuit Art Quarterly* 21 no. 1 (2006): 12–20.

Gustavison, Susan. *Northern Rock: Contemporary Inuit Stone Sculpture*. Kleinberg, ON: McMichael Canadian Art Collection, 1999.

Hall, Alex M. *Discovering Eden: A Lifetime of Paddling Arctic Rivers*. Toronto: Key Porter Books Limited, 2003.

Hallendy, Norman. *Inuksuit: Silent Messengers of the Arctic*. Seattle: University of Washington Press, 2000.

Hankins, Gerald W. *Sunrise over Pangnirtung: The Story of Otto Schaeffer, M.D.* Calgary: The Arctic Institute of North America, 2000.

Harrington, Richard. *The Face of the Arctic*. New York: Henry Schuman, 1952.

Haugen, Einar. *Voyages to Vinland: The First American Saga*. New York: Alfred A Knopf, 1942.

Hessel, Ingo. "Arviat Stone Sculpture: Born of the Struggle with an Uncompromising Medium." *Inuit Art Quarterly* 5, no. 1 (1990): 4–15.

———. *Inuit Art*. London: British Museum Press, 1998.

Houston, C. Stuart, ed. *To the Arctic by Canoe, 1819–1821. The Journal and Paintings of Robert Hood, Midshipman with Franklin*. Montreal: McGill-Queen's University Press, 1994.

Houston, James. *Confessions of an Igloo Dweller: Memories of the Old Arctic*. Boston: Houghton Mifflin Company, 1995.

———. *James Houston's Treasury of Inuit Legends*. Orlando: Harcourt, 2006.

Hummel, Monte, and Justina C. Ray. *Caribou and the North: A Shared Future*. Toronto: Dundurn Press, 2008.

Jenness, Diamond. "Eskimo Art." *Geographical Review* 12, no. 2 (1922): 161–74.

———. *The People of the Twilight*. 1928; repr. Chicago: University of Chicago Press, 1959.

Kane, Elisha Kent. *Arctic Explorations: The Second Grinnell Expedition in Search of Sir John Franklin, 1853, '54, '55*. Philadelphia: Childs & Peterson, 1858.

Kelly, Barry C., et al. "Food Web-Specific Biomagnification of Persistent Organic Pollutants." *Science* 317 (2007): 236–38.

Kent, Rockwell. *N by E*. New York: New York Literary Guild, 1930.

Kerr, Richard A. "Arctic Summer Sea Ice Could Vanish Soon but Not Suddenly." *Science* 323 (2009): 1655.

———. "Geologists Find Vestige of Early Earth—Maybe World's Oldest Rock." *Science* 321 (2008): 1755.

Klevens, R., et al. "Invasive Methicillin-Resitant *Staphylococcus aureus* Infections in the United States." *JAMA* 298 (2007): 1763–71.

Kunnuk, Simeonie, and Janet McGrath. "Lucy Tasseor Tutsuitok: I Portray the Old Way of Life, the Period of Change, and the New Way of Life for the Inuit People." *Inuit Art Quarterly* 13, no. 4 (1998): 20–23.

Lentz, John W. "Inuit Ku: The River of Men." *The Beaver* (Spring 1968): 4–11.

Loomis, Chauncey C. *Weird and Tragic Stories: The Story of Charles Francis Hall, Explorer.* New York: Alfred A. Knopf, 1971.

Lopez, Barry. *Arctic Dreams: Imagination and Desire in a Northern Landscape.* New York: Charles Scribner's Sons, 1986.

Lowenstein, Tom. *Eskimo Poems from Canada and Greenland.* Translated by Tom Lowenstein from material originally collected by Knud Rasmussen. London: Allison and Busby Ltd., 1973.

MacDonald, John. *The Arctic Sky: Inuit Astronomy, Star Lore, and Legend.* Toronto: Royal Ontario Museum, 2000.

Mackinnon, C. S. "Garry Lake, 1948–1958." *Musk-Ox* 38 (1991): 27–44.

Mancall, Peter C. *Fatal Journey: The Final Expedition of Henry Hudson—A Tale of Mutiny and Murder in the Arctic.* New York: Basic Books, 2009.

Mannik, Hattie. *Inuit Nunamiut: Inland Inuit.* Altona, MB: Friesen Corp., 1998.

Marsh, Winifred Petchey. *People of the Willow: The Padlermiut Tribe of the Caribou Eskimo.* Toronto: Oxford University Press, 1976.

McDougall, Anne. "Jenness on Eskimo Art: Documentation with Sympathy and No Pretension." *Inuit Art Quarterly* 7, no. 1 (1992): 22–29.

McGhee, Robert. *The Last Imaginary Place: A Human History of the Arctic World.* New York: Oxford University Press, 2005.

McGoogan, Ken. *Fatal Passage: The Story of John Rae, the Arctic Hero Time Forgot.* New York: Carroll and Graf, 2003.

McGrath, Melanie. *The Long Exile: A Tale of Inuit Betrayal and Survival in the High Arctic.* New York: Alfred A. Knopf, 2006.

McLennan, Jean. Unpublished poem. Personal communication, 2006.

Michell, Marybelle, and Pat Tobin. "Newest Member of Federation." *Inuit Art Quarterly* 14, no. 2 (1999): 18–23.

Millard, Peter. "On Quality in Art: Who Decides?" *Inuit Art Quarterly* 7, no. 3 (1992): 4–14.

Mirsky, Jeannette. *To The Arctic! The Story of Northern Exploration from Earliest Times to the Present.* New York: Alfred A. Knopf, 1948.

Mowat, Farley. *The Desperate People.* Boston: Little, Brown and Company, 1959.

———. *People of the Deer.* Revised ed. Toronto: Seal Books, 1980.

———. *Walking on the Land.* South Royalton, VT: Steerforth Press, 2001.

Nanton, Paul. *Arctic Breakthrough: Franklin's Expeditions, 1819–1847.* Toronto: Clarke, Irwin and Company, Ltd., 1970.

Nasby, Judith. *Irene Avaalaaqiaq: Myth and Reality.* Montreal: McGill-Queen's University Press, 2002.

Paartridge, Taqralik. "Throatsinging: More Than a Game." *Inuit Art Quarterly* 16, no. 4 (2001): 6–10, 28–32.

Palsson, Gisli, ed. *Writing on Ice: The Ethnographic Notebooks of Vilhjalmur Stefansson.* Hanover, NH: University Press of New England, 2001.

Parry, William Edward. *Journal of a Second Voyage for the Discovery of a North-West Passage from the Atlantic to the Pacific, Performed in the Years 1821, 22, 23 in His Majesty's Ships Fury and Hecla.* London: John Murray, 1824.

Pelly, David F., and Christopher C. Hanks, eds. *The Kazan Journey into an Emerging Land.* Yellowknife, NWT: Outcrop, the Northern Publishers, 1991.

Pielou, E. C. *A Naturalist's Guide to the Arctic.* Chicago: University of Chicago Press, 1994.

Post, Eric, et al. "Ecological Dynamics across the Arctic Associated with Recent Climate Change." *Science* 325 (2009): 1355–58.

Powell-Williams, Clive. *Cold Burial: A True Story of Endurance and Disaster.* New York: St. Martin's Press, 2002.

Pryde, Duncan. *Nunanga: Ten Years of Eskimo Life.* New York: Walker and Co, 1971.

Qamukaq, William. "Martina Anoee: 'I Use Inuit Faces as Models.'" *Inuit Art Quarterly* 14, no. 1 (1999): 34–36.

Rasmussen, Knud. *Across Arctic America. Narrative of the Fifth Thule Expedition.* 1927; repr. Anchorage: University of Alaska Press, 1999.

Robinson, Gillian, ed. *Atanarjuat, The Fast Runner.* Igloolik: Coach House Books and Isuma Publishing, 2002.

Roe, John. "I Know My Life Was Happier, Freer." *Kitchener Waterloo Record,* October 1, 1987.

Rowlands, Mark. *The Philosopher and the Wolf: Lessons from the Wild on Love, Death, and Happiness.* New York: Pegasus Books, 2009.

Rowley, Graham W. *Cold Comfort: My Love Affair with the Arctic.* Buffalo, NY: McGill-Queen's University Press, 1996.

Savours, Ann. *The Search for the North West Passage.* New York: St Martin's Press, 1999.

Scherman, Katharine. *Spring on an Arctic Island.* Boston: Little, Brown and Company, 1956.

Screen, James A., and Ian Simmonds. "The Central Role of Diminishing Sea Ice in Recent Arctic Temperature Amplification." *Nature* 464 (2010): 1334–37.

Seton, Ernest Thompson. *The Arctic Prairies: A Canoe Journey of 2,000 Miles in Search of the Caribou, Being the Account of a Voyage to the Region North of Aylmer Lake.* Charleston, SC: BiblioBazaar, 2006.

Shepard, Paul. *Coming Home to the Pleistocene.* Edited by Florence R. Shepard. Washington, DC: Shearwater Books, 1998.

Simon, Alvah. *North into the Night: A Spiritual Odyssey in the Arctic.* New York: Broadway Books, 1998.

Spalding, Alex, with the cooperation and help of Thomas Kusugaq. *Inuktitut: A Multi-Dialectal Outline Dictionary (with an Aivilingmiutaq Base).* Iqaluit: Nunavut Arctic College, 1998.

Steenhoven, Geert van den. "Ennadai Lake People 1955." *The Beaver* (Spring 1968): 12–18.

Stewart, Andrew, T. Max Friesen, Darren Keith, and Lyle Henderson. "Archeology and Oral History of Inuit Land Use on the Kazan River, Nunavut: A Feature-Based Approach." *Arctic* 53 (2000): 260–76.

Swinton, George. *Sculpture of the Inuit.* 3rd ed. Toronto: McClelland and Stewart, Inc., 1972.

Tester, Frank, and Peter Kulchyski. *Tammarniit (Mistakes): Inuit Relocation in the Eastern Arctic, 1939–1963.* Vancouver: UBC Press, 1994.

Verne, Jules. *The Adventures of Captain Hatteras: A New Translation by William Butcher.* New York: Oxford University Press, 2005.

Von Finkenstein, Maria. "Miriam Qiyuk: Variations on a Theme." *Inuit Art Quarterly* 13, no. 4 (1998): 30–33.

———, ed. *Nuvisavik: The Place Where We Weave.* Seattle: University of Washington Press, 2002.

Vors, Liv Solveig, and Mark Stephen Boyce. "Global Declines of Caribou and Reindeer." *Global Change Biology* 15 (2009): 2626–33.

Wakelyn, Leslie. "The Qamanirjuaq Caribou Herd: An Arctic Enigma." June 23, 2010. www.arctic-caribou.com/PDF/qcs.pdf.

Webster, Deborah Kigjugalik. *Harvaqtuurmiut Heritage: The Heritage of the Inuit of the Lower Kazan River.* Yellowknife, NWT: Artisan Press Ltd., 1999.

Wight, Darlene C. "Germaine Arnaktauyok." *Inuit Art Quarterly* 13, no. 2 (1998): 43–45.

GLOSSARY

Inuktitut words other than those marked * are from Alex Spalding, with the cooperation and help of Thomas Kusugaq, *Inuktitut: A Multi-Dialectal Outline Dictionary (with an Aivilingmiutaq Base)* (Iqaluit: Nunavut Arctic College, 1998).

aiviq	walrus
*ajurnarmat	it can't be helped
amaruq	Arctic wolf
*amautik (s); amautiit (pl)	traditional women's parka with hood
angakkuq	shaman
aput	snow
atigi, atigit (pl)	inner parka
atirtaq	polar bear cub when with mother
atirtalik	she-bear with cubs
eleesimayuk	wisdom, knowledge
eyounaqtuq	funny
ikkii	My, it's cold!
Inuk (s); Inuit (pl)	human being, person
Inuktitut	refers to Inuit language
inuksuk (s); inuksuit (pl)	traditional stone beacon usually made of piled stones on some prominent point or hill acting as a guide
inutuqak	elders
iqaluk	fish—specifically, Arctic char
isuma	thought, sense, intelligence, feeling, inspiration, imagination
kamik	boot
matna	thank you (also qujannamiik)
nanuq	polar bear

*natsiq	ringed seal
nikku	dried caribou meat
nuna	land
Nunavut	people of the land
palaugaaq	cooked flour, bannock
paurngaqutit	small berries (such as blueberries)
piqanara	friend
puviksukartuq	hard, dry, screechy sound of sled runners or footsteps over the dry snow
*qallunaaq (s); qallunaat (pl)	white men
qajaq	kayak or man's hunting boat
qamutiik	double-runner sled
qiviu	fine hair (qiviut: underwool of musk-ox)
qujannamiik	thank you
qulliq	stone lamp (burning seal or whale oil)
quvianaktuq	to feel deeply happy; that which causes happiness
quvianaktuvik	heaven, or a place to be happy
siksik	Arctic marmot, Arctic ground squirrel
siku	sea ice
taqqiq	moon
tariuq	the sea, saltwater
tariuraaluk	out of sight of land
tuktu(s); tuktuit (pl)	Arctic caribou
tuqiseeayuk	understanding
ukaliq	Arctic hare
ulu	woman's semicircular or half-moon knife or scraper
umiaq	boat (woman's boat)
uqjuk	bearded seal
umik (s); umit (pl)	facial hair
*umingmak	musk-ox (see above derivation)
urpik	Arctic willow